Suzuki GS850 Fours Owners Workshop Manual

by Martyn Meek
with an additional Chapter on the 1981 to 1988 models
by Jeremy Churchill

Models covered

GS850 GN. UK February to November 1979, US November 1978 to 1979
GS850 GT. UK November 1979 to December 1980, US October 1979 to 1980
GS850 GLT. US October 1979 to 1980
GS850 GX. UK December 1980 to March 1984, US September 1980 to 1981
GS850 GLX. US September 1980 to 1981
GS850 GZ. UK June 1983 to December 1985, US November 1981 to 1982
GS850 GLZ. US September 1981 to 1982
GS850 GD. US October 1982 to 1983
GS850 GLD. US October 1982 to 1983
GS850 GE. UK June 1985 to May 1986
GS850 GG. UK May 1986 to December 1988

All models are fitted with an 843cc engine

ISBN 978 1 85010 571 8

Printed in the UK   (536-1O3)

ABCDE
FGHIJ
KL

2

Haynes Publishing Group
Sparkford Nr Yeovil
Somerset BA22 7JJ England

Haynes Publications, Inc
859 Lawrence Drive
Newbury Park
California 91320 USA

British Library Cataloguing in Publication Data

Meek, Martyn
  Suzuki GS850 Fours owners workshop manual.
  1. Motorcycles. Maintenance & repairs – Amateurs' manuals
  I. Title   II. Churchill, Jeremy. 1954-   III. Series
  629.28'775
  ISBN 1-85010-571-5

Library of Congress Catalog Card Number
88-83913

Acknowledgements

Our thanks are due to P.R. Taylor and Sons of Chippenham who loaned the GS850 GN featured in the photographs throughout this manual, and to Steve Wilson of Blandford who provided the GS850 GG featured on the front cover.

Brian Horsfall supervised and assisted with the dismantling of the machine and also the rebuilding sequences, and devised various ingenious methods for overcoming the lack of service tools. Les Brazier arranged and took the photographs and Mansur Darlington edited the text.

We should also like to thank the Avon Rubber Company who kindly supplied us with information and advice about tyre fitting, and NGK Spark Plugs (UK) Limited, for information and photographs relating to sparking plug conditions.

About this manual

The purpose of this manual is to present the owner with a concise and graphic guide which will enable him to tackle any operation from basic routine maintenance to a major overhaul. It has been assumed that any work would be undertaken without the luxury of a well-equipped workshop and a range of manufacturer's service tools.

To this end, the machine featured in the manual was stripped and rebuilt in our own workshop, by a team comprising a mechanic, a photographer and the author. The resulting photographic sequence depicts events as they took place, the hands shown being those of the author and the mechanic.

The use of specialised, and expensive, service tools was avoided unless their use was considered to be essential due to risk of breakage or injury. There is usually some way of improving a method of removing a stubborn component, provided that a suitable degree of care is exercised.

The author learnt his motorcycle mechanics over a number of years, faced with the same difficulties and using similar facilities to those encountered by most owners. It is hoped that this practical experience can be passed on through the pages of this manual.

Where possible, a well-used example of the machine is chosen for the workshop project, as this highlights any areas which might be particularly prone to giving rise to problems. In this way, any such difficulties are encountered and resolved before the text is written, and the techniques used to deal with them can be incorporated in the relevant sections. Armed with a working knowledge of the machine, the author undertakes a considerable amount of research in order that the maximum amount of data can be included in this manual.

Each Chapter is divided into numbered sections. Within these sections are numbered paragraphs. Cross reference throughout the manual is quite straightforward and logical. When reference is made 'See Section 6.10' it means Section 6, paragraph 10 in the same Chapter. If another Chapter were intended the reference would read, for example, 'See Chapter 2, Section 6.10'. All the photographs are captioned with a section/paragraph number to which they refer and are relevant to the Chapter text adjacent.

Figures (usually line illustrations) appear in a logical but numerical order, within a given Chapter. Fig. 1.1 therefore refers to the first figure in Chapter 1.

Left-hand and right-hand descriptions of the machines and their components refer to the left and right of a given machine when the rider is seated normally.

Motorcycle manufacturers continually make changes to specifications and recommendations, and these, when notified, are incorporated into our manuals at the earliest opportunity.

Whilst every care is taken to ensure that the information in this manual is correct no liability can be accepted by the author or publishers for loss, damage or injury caused by any errors in or omissions from the information given.

Contents

Right-hand view of Suzuki GS 850 GN – UK model

Left-hand view of Suzuki GS 850 GN – UK model

Introduction to the Suzuki GS 850 models

Although the Suzuki Motor Company Limited commenced manufacturing motorcycles as early as 1936, it was not until 1963 that their machines were first imported into the UK. The first of the twin cylinder models, the T10, became available during 1964, and it was immediately obvious that this model would be well-received by holders of a provisional driving licence, who are restricted to an engine capacity of 250 cc. Its popularity was mostly due to its impressive performance, in terms of both speed and fuel economy, and its advanced specification, including an electric starter and a hydraulic rear brake.

Being very similar to, and indeed mechanically based upon, the existing four-cylinder GS models, the GS 850 has inherited many of the more useful features and more endearing characteristics of the other models. The US market GS 850 models have made use of the excellent air-assisted front forks of the fine handling GS 1000 models, whilst at the rear of a sturdy frame, the rear suspension units incorporate a sophisticated variable rebound damping facility. The frame design and rear suspension units are to be found on all the GS 850 models. These factors combine to ensure that, although an extremely heavy motorcycle, the GS 850 handles safely, responsively and predictably. The majority of the weight increase that the GS 850 suffers over its stablemates (it is substantially heavier than the GS 1000 for instance) stems from the difficulty in creating a lightweight shaft final drive system. Excess weight is a seemingly inevitable penalty of a trouble free final drive system for motorcycle application.

The initial GS 850 GN model was introduced in 1979 in both the USA and UK. For the 1980 model year, the GT model was introduced to both US and UK markets. Whilst being basically the same machine, the GT has received detail refinements and modifications in several areas, to uprate its specification. Detail refinements of most significance are to be found on the handlebar (new control levers, switchgear and brake fluid reservoir design), in the headlamp (a new unit utilising a quartz halogen bulb), and behind the left-hand side panel, where a redesigned fuse box incorporates the fused output terminal; it was a separate unit previously. More major modifications incorporated on the GT model include a set of constant vacuum carburettors, a redesigned exhaust system incorporating a pre-silencer chamber, slotted brake discs with different calipers, and a transistorised ignition system.

A second variant, the GLT model has now been introduced. The 'L' variant is styled in the 'semi-Chopper' mould; a style becoming increasingly popular and now involving all the four large Japanese manufacturers. The GLT model is only available in the USA.

To assist owners in identifying their machine exactly, given below are the initial frame numbers with which each model's production run commenced, together with the approximate dates of import. Note that the latter may not necessarily coincide with the machine's date of registration or sale.

GS 850 GN	GS850–100001 on	UK Feb to Nov 1979
		US Nov 1978 to 1979
GS 850 GT	GS850–116014 on	UK Nov 1979 to Dec 1980
		US Oct 1979 to 1980
GS 850 GLT	GS850–700001 on	US Oct 1979 to 1980

Details of the 1981 to 1988 models will be found in Chapter 7 of this manual.

Model dimensions and weights

	GS 850 GN	GS 850 GT	GS 850 GLT
Overall length	2230 mm (87.8 in)	2230 mm (87.8 in)	2285 mm (90.0 in)
Overall width	865 mm (34.1 in)	865 mm (34.1 in)	900 mm (35.4 in)
Overall height	1190 mm (46.9 in)	1190 mm (46.9 in)	1230 mm (48.4 in)
Wheelbase	1490 mm (58.7 in)	1500 mm (59.1 in)	1510 mm (59.4 in)
Ground clearance	160 mm (6.3 in)	160 mm (6.3 in)	150 mm (5.9 in)
Dry weight	253 kg 558 lbs	253 kg 558 lbs	245 kg 540 lbs

Ordering spare parts

When ordering spare parts for any Suzuki, it is advisable to deal direct with an official Suzuki agent who should be able to supply most of the parts ex stock. Parts cannot be obtained from Suzuki direct and all orders must be routed via an approved agent even if the parts required are not held in stock. Always, quote the engine and frame numbers in full, especially if parts are required for earlier models.

The frame and engine numbers are stamped on a Manufacturer's Plate riveted to the steering head on the left-hand side. The frame number is also stamped on the frame itself on the right-hand side of the steering head. The engine number is stamped on the upper crankcase.

Use only genuine Suzuki spares. Some pattern parts are available that are made in Japan and may be packed in similar looking packages. They should only be used if genuine parts are hard to obtain or in an emergency, for they do not normally last as long as genuine parts, even though there may be a price advantage.

Some of the more expendable parts such as spark plugs, bulbs, tyres, oils and greases etc., can be obtained from accessory shops and motor factors, who have convenient opening hours, and can often be found not far from home. It is also possible to obtain parts on a Mail Order basis from a number of specialists who advertise regularly in the motorcycle magazines.

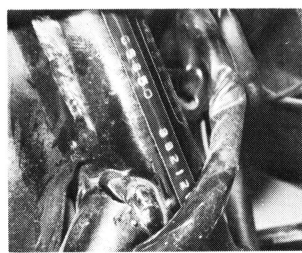

Location of frame number

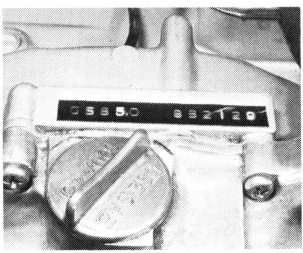

Location of engine number

Safety first!

Professional motor mechanics are trained in safe working procedures. However enthusiastic you may be about getting on with the job in hand, do take the time to ensure that your safety is not put at risk. A moment's lack of attention can result in an accident, as can failure to observe certain elementary precautions.

There will always be new ways of having accidents, and the following points do not pretend to be a comprehensive list of all dangers; they are intended rather to make you aware of the risks and to encourage a safety-conscious approach to all work you carry out on your vehicle.

Essential DOs and DON'Ts

DON'T start the engine without first ascertaining that the transmission is in neutral.

DON'T suddenly remove the filler cap from a hot cooling system – cover it with a cloth and release the pressure gradually first, or you may get scalded by escaping coolant.

DON'T attempt to drain oil until you are sure it has cooled sufficiently to avoid scalding you.

DON'T grasp any part of the engine, exhaust or silencer without first ascertaining that it is sufficiently cool to avoid burning you.

DON'T allow brake fluid or antifreeze to contact the machine's paintwork or plastic components.

DON'T syphon toxic liquids such as fuel, brake fluid or antifreeze by mouth, or allow them to remain on your skin.

DON'T inhale dust – it may be injurious to health (see *Asbestos* heading).

DON'T allow any spilt oil or grease to remain on the floor – wipe it up straight away, before someone slips on it.

DON'T use ill-fitting spanners or other tools which may slip and cause injury.

DON'T attempt to lift a heavy component which may be beyond your capability – get assistance.

DON'T rush to finish a job, or take unverified short cuts.

DON'T allow children or animals in or around an unattended vehicle.

DON'T inflate a tyre to a pressure above the recommended maximum. Apart from overstressing the carcase and wheel rim, in extreme cases the tyre may blow off forcibly.

DO ensure that the machine is supported securely at all times. This is especially important when the machine is blocked up to aid wheel or fork removal.

DO take care when attempting to slacken a stubborn nut or bolt. It is generally better to pull on a spanner, rather than push, so that if slippage occurs you fall away from the machine rather than on to it.

DO wear eye protection when using power tools such as drill, sander, bench grinder etc.

DO use a barrier cream on your hands prior to undertaking dirty jobs – it will protect your skin from infection as well as making the dirt easier to remove afterwards; but make sure your hands aren't left slippery. Note that long-term contact with used engine oil can be a health hazard.

DO keep loose clothing (cuffs, tie etc) and long hair well out of the way of moving mechanical parts.

DO remove rings, wristwatch etc, before working on the vehicle – especially the electrical system.

DO keep your work area tidy – it is only too easy to fall over articles left lying around.

DO exercise caution when compressing springs for removal or installation. Ensure that the tension is applied and released in a controlled manner, using suitable tools which preclude the possibility of the spring escaping violently.

DO ensure that any lifting tackle used has a safe working load rating adequate for the job.

DO get someone to check periodically that all is well, when working alone on the vehicle.

DO carry out work in a logical sequence and check that everything is correctly assembled and tightened afterwards.

DO remember that your vehicle's safety affects that of yourself and others. If in doubt on any point, get specialist advice.

IF, in spite of following these precautions, you are unfortunate enough to injure yourself, seek medical attention as soon as possible.

Asbestos

Certain friction, insulating, sealing, and other products – such as brake linings, clutch linings, gaskets, etc – contain asbestos. *Extreme care must be taken to avoid inhalation of dust from such products since it is hazardous to health.* If in doubt, assume that they *do* contain asbestos.

Fire

Remember at all times that petrol (gasoline) is highly flammable. Never smoke, or have any kind of naked flame around, when working on the vehicle. But the risk does not end there – a spark caused by an electrical short-circuit, by two metal surfaces contacting each other, by careless use of tools, or even by static electricity built up in your body under certain conditions, can ignite petrol vapour, which in a confined space is highly explosive.

Always disconnect the battery earth (ground) terminal before working on any part of the fuel or electrical system, and never risk spilling fuel on to a hot engine or exhaust.

It is recommended that a fire extinguisher of a type suitable for fuel and electrical fires is kept handy in the garage or workplace at all times. Never try to extinguish a fuel or electrical fire with water.

Note: *Any reference to a 'torch' appearing in this manual should always be taken to mean a hand-held battery-operated electric lamp or flashlight. It does **not** mean a welding/gas torch or blowlamp.*

Fumes

Certain fumes are highly toxic and can quickly cause unconsciousness and even death if inhaled to any extent. Petrol (gasoline) vapour comes into this category, as do the vapours from certain solvents such as trichloroethylene. Any draining or pouring of such volatile fluids should be done in a well ventilated area.

When using cleaning fluids and solvents, read the instructions carefully. Never use materials from unmarked containers – they may give off poisonous vapours.

Never run the engine of a motor vehicle in an enclosed space such as a garage. Exhaust fumes contain carbon monoxide which is extremely poisonous; if you need to run the engine, always do so in the open air or at least have the rear of the vehicle outside the workplace.

The battery

Never cause a spark, or allow a naked light, near the vehicle's battery. It will normally be giving off a certain amount of hydrogen gas, which is highly explosive.

Always disconnect the battery earth (ground) terminal before working on the fuel or electrical systems.

If possible, loosen the filler plugs or cover when charging the battery from an external source. Do not charge at an excessive rate or the battery may burst.

Take care when topping up and when carrying the battery. The acid electrolyte, even when diluted, is very corrosive and should not be allowed to contact the eyes or skin.

If you ever need to prepare electrolyte yourself, always add the acid slowly to the water, and never the other way round. Protect against splashes by wearing rubber gloves and goggles.

Mains electricity

When using an electric power tool, inspection light etc which works from the mains, always ensure that the appliance is correctly connected to its plug and that, where necessary, it is properly earthed (grounded). Do not use such appliances in damp conditions and, again, beware of creating a spark or applying excessive heat in the vicinity of fuel or fuel vapour.

Ignition HT voltage

A severe electric shock can result from touching certain parts of the ignition system, such as the HT leads, when the engine is running or being cranked, particularly if components are damp or the insulation is defective. Where an electronic ignition system is fitted, the HT voltage is much higher and could prove fatal.

Conversion factors

Length (distance)

	X		=		X		=	
Inches (in)	X	25.4	=	Millimetres (mm)	X	0.0394	=	Inches (in)
Feet (ft)	X	0.305	=	Metres (m)	X	3.281	=	Feet (ft)
Miles	X	1.609	=	Kilometres (km)	X	0.621	=	Miles

Volume (capacity)

	X		=		X		=	
Cubic inches (cu in; in³)	X	16.387	=	Cubic centimetres (cc; cm³)	X	0.061	=	Cubic inches (cu in; in³)
Imperial pints (Imp pt)	X	0.568	=	Litres (l)	X	1.76	=	Imperial pints (Imp pt)
Imperial quarts (Imp qt)	X	1.137	=	Litres (l)	X	0.88	=	Imperial quarts (Imp qt)
Imperial quarts (Imp qt)	X	1.201	=	US quarts (US qt)	X	0.833	=	Imperial quarts (Imp qt)
US quarts (US qt)	X	0.946	=	Litres (l)	X	1.057	=	US quarts (US qt)
Imperial gallons (Imp gal)	X	4.546	=	Litres (l)	X	0.22	=	Imperial gallons (Imp gal)
Imperial gallons (Imp gal)	X	1.201	=	US gallons (US gal)	X	0.833	=	Imperial gallons (Imp gal)
US gallons (US gal)	X	3.785	=	Litres (l)	X	0.264	=	US gallons (US gal)

Mass (weight)

	X		=		X		=	
Ounces (oz)	X	28.35	=	Grams (g)	X	0.035	=	Ounces (oz)
Pounds (lb)	X	0.454	=	Kilograms (kg)	X	2.205	=	Pounds (lb)

Force

	X		=		X		=	
Ounces-force (ozf; oz)	X	0.278	=	Newtons (N)	X	3.6	=	Ounces-force (ozf; oz)
Pounds-force (lbf; lb)	X	4.448	=	Newtons (N)	X	0.225	=	Pounds-force (lbf; lb)
Newtons (N)	X	0.1	=	Kilograms-force (kgf; kg)	X	9.81	=	Newtons (N)

Pressure

	X		=		X		=	
Pounds-force per square inch (psi; lbf/in²; lb/in²)	X	0.070	=	Kilograms-force per square centimetre (kgf/cm²; kg/cm²)	X	14.223	=	Pounds-force per square inch (psi; lbf/in²; lb/in²)
Pounds-force per square inch (psi; lbf/in²; lb/in²)	X	0.068	=	Atmospheres (atm)	X	14.696	=	Pounds-force per square inch (psi; lbf/in²; lb/in²)
Pounds-force per square inch (psi; lbf/in²; lb/in²)	X	0.069	=	Bars	X	14.5	=	Pounds-force per square inch (psi; lbf/in²; lb/in²)
Pounds-force per square inch (psi; lbf/in²; lb/in²)	X	6.895	=	Kilopascals (kPa)	X	0.145	=	Pounds-force per square inch (psi; lbf/in²; lb/in²)
Kilopascals (kPa)	X	0.01	=	Kilograms-force per square centimetre (kgf/cm²; kg/cm²)	X	98.1	=	Kilopascals (kPa)
Millibar (mbar)	X	100	=	Pascals (Pa)	X	0.01	=	Millibar (mbar)
Millibar (mbar)	X	0.0145	=	Pounds-force per square inch (psi; lbf/in²; lb/in²)	X	68.947	=	Millibar (mbar)
Millibar (mbar)	X	0.75	=	Millimetres of mercury (mmHg)	X	1.333	=	Millibar (mbar)
Millibar (mbar)	X	0.401	=	Inches of water (inH₂O)	X	2.491	=	Millibar (mbar)
Millimetres of mercury (mmHg)	X	0.535	=	Inches of water (inH₂O)	X	1.868	=	Millimetres of mercury (mmHg)
Inches of water (inH₂O)	X	0.036	=	Pounds-force per square inch (psi; lbf/in²; lb/in²)	X	27.68	=	Inches of water (inH₂O)

Torque (moment of force)

	X		=		X		=	
Pounds-force inches (lbf in; lb in)	X	1.152	=	Kilograms-force centimetre (kgf cm; kg cm)	X	0.868	=	Pounds-force inches (lbf in; lb in)
Pounds-force inches (lbf in; lb in)	X	0.113	=	Newton metres (Nm)	X	8.85	=	Pounds-force inches (lbf in; lb in)
Pounds-force inches (lbf in; lb in)	X	0.083	=	Pounds-force feet (lbf ft; lb ft)	X	12	=	Pounds-force inches (lbf in; lb in)
Pounds-force feet (lbf ft; lb ft)	X	0.138	=	Kilograms-force metres (kgf m; kg m)	X	7.233	=	Pounds-force feet (lbf ft; lb ft)
Pounds-force feet (lbf ft; lb ft)	X	1.356	=	Newton metres (Nm)	X	0.738	=	Pounds-force feet (lbf ft; lb ft)
Newton metres (Nm)	X	0.102	=	Kilograms-force metres (kgf m; kg m)	X	9.804	=	Newton metres (Nm)

Power

	X		=		X		=	
Horsepower (hp)	X	745.7	=	Watts (W)	X	0.0013	=	Horsepower (hp)

Velocity (speed)

	X		=		X		=	
Miles per hour (miles/hr; mph)	X	1.609	=	Kilometres per hour (km/hr; kph)	X	0.621	=	Miles per hour (miles/hr; mph)

Fuel consumption*

	X		=		X		=	
Miles per gallon, Imperial (mpg)	X	0.354	=	Kilometres per litre (km/l)	X	2.825	=	Miles per gallon, Imperial (mpg)
Miles per gallon, US (mpg)	X	0.425	=	Kilometres per litre (km/l)	X	2.352	=	Miles per gallon, US (mpg)

Temperature

Degrees Fahrenheit = (°C x 1.8) + 32

Degrees Celsius (Degrees Centigrade; °C) = (°F - 32) x 0.56

*It is common practice to convert from miles per gallon (mpg) to litres/100 kilometres (l/100km), where mpg (Imperial) x l/100 km = 282 and mpg (US) x l/100 km = 235

Routine maintenance

Refer to Chapter 7 for information relating to the 1981 to 1988 models

Periodic routine maintenance is a continuous process that commences immediately the machine is used and continues until the machine is no longer fit for service. It must be carried out at specified mileage recordings or on a calendar basis if the machine is not used regularly, whichever is the sooner. Maintenance should be regarded as an insurance policy, to help keep the machine in the peak of condition and to ensure long trouble-free service. It has the additional benefit of giving early warning of any faults that may develop and will act as a safety check to the obvious advantage of both rider and machine alike.

The various maintenance tasks are described under their respective mileage and calendar headings. Accompanying photos or diagrams are provided, where necessary. It should be remembered that the interval between the various maintenance tasks serves only as a guide. As the machine gets older, is driven hard, or is used under particularly adverse conditions, it is advisable to reduce the period between each check.

For ease of reference each service operation is described in detail under the relevant heading. However, if further general information is required it can be found within the manual in the relevant Chapter.

Although no special tools are required for routine maintenance, a good selection of general workshop tools are essential. Included in the tools must be a range of metric ring or combination spanners, a selection of crosshead screwdrivers, and two pairs of circlip pliers, one external opening and the other internal opening. Additionally, owing to the extreme tightness of most casing screws on Japanese machines, an impact screwdriver, together with a choice of large or small cross-head screw bits, is absolutely indispensable. This is particularly so if the engine has not been dismantled since leaving the factory.

Weekly or every 250 miles (400 km)

1 Tyre pressures

Check the tyre pressures with a pressure gauge that is known to be accurate. Always check the pressures when the tyres are cold. If the tyres are checked after the machine has travelled a number of miles, the tyres will have become hot and consequently the pressure will have increased, possibly as much as 8 psi. A false reading will therefore always result.

Tyre pressures: *Solo*   *Pillion*
   Front tyre   *25 psi (1.75 kg/cm²)*   *25 psi (1.75 kg/cm²)*
   Rear tyre   *28 psi (2.00 kg/cm²)*   *32 psi (2.25 kg/cm²)*
For continuous high speed riding, the pressures should be increased to:
             Solo   *Pillion*
   Front tyre   *28 psi (2.00 kg/cm²)*   *28 psi (2.00 kg/cm²)*
   Rear tyre   *32 psi (2.25 kg/cm²)*   *40 psi (2.80 kg/cm²)*

2 Engine/gearbox oil level

Place the machine on the centre stand and by viewing the sight-glass in the primary drive casing, check that the engine/transmission oil level is between the two level marks. The machine must be upright because even a slight lean will give a false reading. If necessary, replenish the engine with the correct quantity of SAE 10W/40 engine oil. The filler cap is situated in the top of the primary drive cover.

Check tyre pressures with tyres cold

Check engine/transmission oil level through sight-glass

3 Control cable lubricating

Apply a few drops of motor oil to the exposed inner portion of each control cable. This will prevent drying-up of the cables between the more thorough lubrication that should be carried out during the 4000 mile/4 monthly service.

4 Safety check

Give the machine a close visual inspection, checking for loose nuts and fittings, frayed control cables, etc. Check the tyres for damage, especially splitting of the sidewalls. Remove any stones or other objects caught between the treads. This is particularly important on the front tyre, where rapid deflation due to the penetration of the inner tube will almost certainly cause total loss of control. When checking the tyres for damage, they should be examined for tread depth in view of both the legal and safety aspects. It is vital to keep the tread depth within the legal limits of 1 mm of depth over at least three quarters of the tread breadth and around the entire circumference. Many riders, however, consider nearer 2 mm to be the limit for secure road-holding, traction and braking, specially when riding in adverse weather and road conditions. Suzuki recommend replacement of the front tyre when there is 1.6 mm (0.06 in) of tread remaining, and replacement of the rear tyre when wear has reached the 2.0 mm (0.08 in) level.

5 Legal check

Ensure that the lights, horn, and flashing indicators all function correctly, and check that the speedometer is working and is reasonably accurate; the law requires it to be accurate within 10 per cent in the UK.

Monthly or every 1000 miles (1600 km)

Complete the tasks listed under the weekly/250 mile heading and then carry out the following checks:

1 Hydraulic fluid level

Check the level of the hydraulic fluid in the front brake master cylinder reservoir, on the handlebars, and also in the rear brake reservoir, behind the right-hand side cover. The level can be seen through the transparent reservoir and should be between the upper and lower level marks. Ensure that the handlebars are in the central position when a level reading is taken from the front reservoir, and also when the cap and diaphragm are removed. During normal service, it is unlikely that the hydraulic fluid level will drop dramatically unless a leak has developed in the system. If this occurs, the fault should be remedied **AT ONCE**. The level will fall slowly as the brake pads wear, and the fluid deficiency should be corrected when required. Always use a hydraulic fluid of DOT 3 (USA) or SAE J1703 specification and if possible do not mix different types of fluid, even if the specifications appear the same. This will prevent the possibility of two incompatible fluids being mixed and the resultant chemical reactions damaging the seals.

If the level in either reservoir has been allowed to fall below the specified limit, and air has entered the system, the brake in question must be bled, as described in Chapter 5, Section 7.

2 Battery electrolyte level

Maintenance of the battery fitted to the GS 850 range is normally limited to keeping the electrolyte level correct. The transparent plastic case of the battery case permits the level of the electrolyte to be checked when the right-hand frame side cover is detached. The battery is retained in a metal carrier below the tool tray, which in turn, is situated below the dualseat. To replenish the battery is a simple task, not requiring the battery to be removed from the machine. Unlock and raise the dualseat or unlock and detach the dualseat on the GLT model, and remove the tool kit and its tray, to expose the top of the battery.

The electrolyte solution should be between the upper and lower level lines. If the electrolyte solution is low it should be replenished, using distilled water. The lead plates and their separators can be seen through the transparent case, a further guide to the general condition of the battery. Note that a thin vent pipe is fitted to the battery case; this must always be kept free from blockage, and must be routed so as to avoid any sharp turns or kinks.

Unless acid is spilt, as may occur if the machine falls over, the electrolyte should always be topped up with distilled water, to restore the correct level. If acid is spilt on any part of the machine, it should be neutralised with an alkali such as washing soda and washed away with plenty of water, otherwise serious corrosion will occur. Top up with sulphuric acid of the correct specific gravity 1.260 – 1.280 only when spillage has occured.

Top up brake fluid as necessary to maintain correct level

Maintain electrolyte level between upper and lower lines

Complete the checks listed under weekly/500 mile and monthly/1000 mile headings and then carry out the following tasks:

1   Engine/gearbox oil change
The engine oil should be changed with the engine at its normal operating temperature, preferably after a run. This ensures that the oil is relatively thin and will drain more quickly and completely.

Drain the engine oil by removing the drain plug from the underside of the sump and the filler cap from the top of the primary drive cover. Ensure that a container of sufficient size is placed below the engine to catch the oil. When all the oil has drained off, refit and tighten the drain plug after checking the sealing washer is in good condition.

Replenish the engine with approximately 2.8 litres (5.9/4.9 US/Imp pint) of SAE 10W/40 motor oil. Allow the oil to settle, and run the engine briefly at idling speed. Then recheck the oil level, by means of the sight-glass, and, if necessary, add more lubricant. The oil should be roughly three-quarters of the way up the sight-glass.

2   Secondary bevel gear unit and final drive gear case oil level check
The oil level in both the secondary gear casing, which contains the secondary drive and driven bevel gear units, and the final drive gear case (at the rear wheel) should be checked at regular intervals. To check the oil level in the secondary gear casing, remove the oil level screw from the rear of the casing, on the left-hand side of the engine. The screw is clearly marked as to its function, with 'oil level' stamped on the casing. If a small dribble of oil emerges as the screw is withdrawn, the oil level is correct. Check the oil level in the rear wheel final drive gear case by visual inspection. Remove the oil filler plug from the unit, and inspect the oil level within the case. If the oil level is roughly flush with the base of the filler orifice, the level is correct. In normal operation, the oil level in these secondary drive gear and final drive gear casings will vary very little. If replenishment is necessary, use a Hypoid gear oil of SAE 90 viscosity and conforming to GL-5 specification (USA).

Carry out the tasks described in the weekly, monthly and two monthly sections and then carry out the following:

1   Oil filter renewal
The oil filter element should be renewed at every second oil change. After draining the engine oil remove the three domed nuts from the oil filter chamber cover. Ensure a container is placed below the chamber to catch the escaping oil. The cover is under tension from the filter locating spring and so may fly off if care is not taken. Lift out the spring and oil filter element. No attempt should be made to clean the old filter; it must be discarded and a new component fitted. Clean the filter chamber before inserting the new element which should be fitted with the rubber seal end facing inwards. Check the condition of the chamber cover sealing ring before replacing the cover.

Note that when the engine oil and filter are both replaced at the same service interval, 3.6 litres (7.6/6.4 US/Imp pint) of SAE 10W/40 engine oil will be necessary to replenish fully the sump.

Check secondary drive casing oil level by removing oil level screw

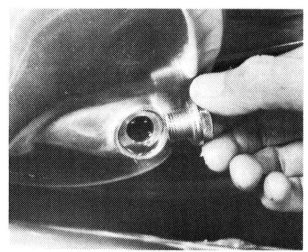
Check final drive gear case oil level by removing filler bolt

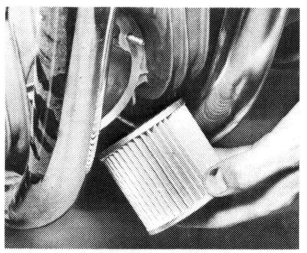
Always use a *new* element at each change

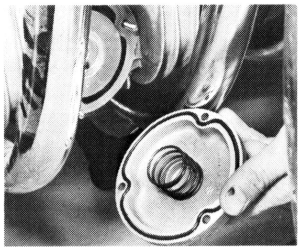

Check sealing ring in chamber cover plate before refitting

2 Cleaning and adjusting the contact breaker points – GS 850 GN model only

To gain access to the contact breaker assembly it is necessary to remove the engine right-hand cover which is retained by three screws.

Before adjusting the points, examine each set for burning or pitting. Deposits due to arcing can be removed while the contact breaker unit is in situ on the machine, using a very fine Swiss file or emery paper (No. 400) backed by a thin strip of tin. If the pitting or burning is excessive, the contact breaker unit in question must be removed for points dressing or renewal as described in Chapter 3.

The points are marked '1.4' and '2.3' adjacent to the relevant contact set, indicating the pair of cylinders they serve.

Set the gap of points marked 1.4 first. Turn the crankshaft using a spanner on the large hexagonal washer securing the cam, until these points are fully open. Measure the gap with a feeler gauge, and adjust if necessary. The standard gap is 0.3 – 0.4 mm (0.012 – 0.016 inch).

If the gap requires adjustment, slacken slightly the slotted screw which secures the fixed contact. A screwdriver should be engaged btween the slot in the fixed contact, and the two pins on the contact breaker back plate; by turning the screwdriver the gap may be opened or closed. Tighten the screw and recheck the gap.

Turn the crankshaft so that the points marked 2.3 are fully open, and repeat the procedure above. Do not slacken the two screws which secure the 2.3 contact set base plate to the main base plate; this will upset the timing. It is important that both points should be set to the same gap, as the gap determines the moment when the contact opens, and thus the ignition timing.

3 Checking and resetting the ignition timing – GS 850 GN model only

Whenever the contact breaker unit receives attention, the ignition timing should be checked as a matter of course and adjusted, if necessary.

Apply a spanner to the engine turning hexagon and turn the engine in a forward direction, whilst viewing the automatic timing unit through the inspection aperture in the contact breaker stator plate. It will be seen that there is a set of three scribed lines on each side of the ATU.

Commence ignition timing on the left-hand contact breaker set, which controls cylinders No. 1 and 4. To determine the point at which the points open, connect a 12 volt bulb between the moving point and a suitable earth point on the engine. With the ignition turned on, the bulb will illuminate when the points are open. Rotate the engine until the 'F1-4' mark on the ATU is

in **exact** alignment with the index pointer mark on the plate fitted to the rear of the stator plate. If the ignition is correct, the points should be on the verge of opening when this position is reached. This will be indicated by the flickering of the bulb. To adjust the ignition timing on No. 1 and 4 cylinders, slacken the three screws which pass through the elongated holes in the stator plate periphery. Rotate the plate until the light flickers and then tighten the screw. Turn the engine backwards 90° and then forwards again to check the setting.

Check the ignition timing on No. 2 and 3 cylinders in a similar manner, by connecting the bulb to the right-hand contact breaker set and referring to the F2-3 mark on the ATU. If the timing is incorrect, slacken the two screws which hold the right-hand contact breaker assembly mounting plate to the main stator plate. Move the plate to the correct position and tighten the screw. Recheck the timing on No. 2 and 3 cylinder.

The ignition timing may also be checked by the use of a stroboscope when the engine is running.

Check and adjust the 1 – 4 contact breaker first and then the 2 – 3 contact breaker. At 1500 rpm and below, the F mark should align with the index mark. Full advance should be reached at 2500 rpm, when the unmarked advance line to the right of the F mark should be in line with the index mark. Adjustment should be made in the same manner as described for manual ignition timing.

Before replacing the contact breaker cover, apply a small quantity of light oil or grease to the cam lubricating wick. Do not overlubricate, as excess oil may find its way onto the points, causing ignition failure.

Apply lubricant to felt cam wick (GN model only)

4 Checking the ignition timing – GT and GLT models

The ignition system fitted to the GT and GLT models is of the fully transistorised type, having no mechanically wearing parts other than the automatic timing unit. Because of this, unless component failure occurs, periodic checking of the ignition timing should not be necessary.

5 Valve clearance checking and adjustment

To gain access to the camshafts and cam followers the petrol tank must be removed as described in Chapter 2, Section 2, and the camshaft cover detached. After removal of the cover bolts, the seal between the cover and the gasket may be broken by the judicious use of a soft-headed mallet. Strike only those parts of the cover which are well supported by lugs.

Unscrew the spark plugs and remove the contact breaker cover, GN model, or the signal generating unit cover on the GT/GLT models, from the right-hand side of the engine. The clearance between each cam and cam follower must be checked and if necessary adjusted by removal of the existing adjuster pad and replacement of a pad of suitable thickness. Make the clearance check and adjustment of each valve in sequence and then continue with the next valve. As shown in the accompanying diagram, both operations should be carried out with the cam lobe in question in one of two alternative positions. Rotate the engine in a forward direction by means of the engine turning hexagon on the contact breaker cam end. Use only the 19 mm hexagon for this purpose.

Using a feeler gauge, determine and record the clearance at the first valve. If the clearance is incorrect, not being within the range of 0.03 – 0.08 mm (0.001 – 0.003 in), the adjuster pad must be removed and replaced by one of suitable thickness. A special tool is available (Suzuki part No. 09916 – 64510) which may be pushed between the camshaft adjacent to the cam lobe and the raised edge of the cam follower, to allow removal of the shim. If the special tool is not available, a simple substitute may be fabricated from a portion of steel plate.

The final form of the tool which has a handle about 6 inches long, can be seen in the accompanying photograph.

The Suzuki tool may be pushed into position, depressing the cam follower and securing it in a depressed position, in one operation. Where a home-made tool is used, the cam follower may be depressed using a suitable lever placed between the adjuster pad and the cam lobe. The tool may then be inserted to secure the cam follower whilst the adjuster pad is removed.

Before installing either type of tool, rotate the cam follower so that the slot in the raised edge is not obscured by the camshaft. Insert a small screwdriver through the slot to displace the adjuster shim.

Adjustment pads are available in 20 sizes ranging from 2.15 mm to 3.10 mm, in increments of 0.05 mm. Each pad is identified by a three digit number etched on the reverse face. The number (eg. 235) indicates that the pad thickness is 2.35 mm thick. Having taken clearance measurements, refer to the accompanying table and the example given to the right of that table to select the correct pads.

Although the adjuster pads are available as a complete set their price is prohibitive. It is suggested that pads are purchased individually, after an accurate assessment of requirement has been made. It is possible that some Suzuki service agents will be prepared to exchange needed pads for others of the correct size, providing that the original pads are not worn.

Before installing a replacement pad, lubricate both sides thoroughly with engine oil. Always fit the pad with the identification number downwards, so that it does not become obliterated by the action of the cam. After fitting new adjuster pads, rotate the engine forwards a number of times and then recheck the clearances to verify that no errors have occurred.

Before refitting the cam cover, together with a new gasket, lubricate the camshafts with copious quantities of clean engine oil.

Check valve clearances using a feeler gauge

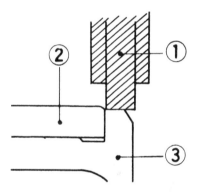

Shim removal using a tappet depressing tool

1   Tappet depressing tool
2   Shim
3   Tappet

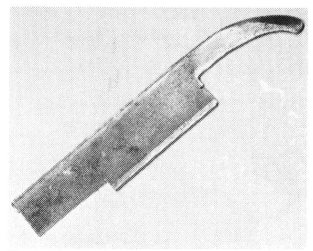

Home-made cam follower depression tool

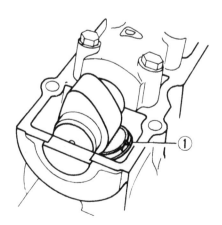

Tappet shim removal position

P/N SUFFIX →	45000	45001	45002	45003	45004	45005	45006	45007	45008	45009	45010	45011	45012	45013	45014	45015	45016	45017	45018	45019
Tappet Clearance (mm)	*PRESENT SHIM SIZE — mm*																			
(present shim size)	2.15	2.20	2.25	2.30	2.35	2.40	2.45	2.50	2.55	2.60	2.65	2.70	2.75	2.80	2.85	2.90	2.95	3.00	3.05	3.10
0.00~0.02		2.15	2.20	2.25	2.30	2.35	2.40	2.45	2.50	2.55	2.60	2.65	2.70	2.75	2.80	2.85	2.90	2.95	3.00	3.05
0.03~0.08	SPECIFIED CLEARANCE/NO. ADJUSTMENT REQUIRED																			
0.09~0.13	2.20	2.25	2.30	2.35	2.40	2.45	2.50	2.55	2.60	2.65	2.70	2.75	2.80	2.85	2.90	2.95	3.00	3.05	3.10	
0.14~0.18	2.25	2.30	2.35	2.40	2.45	2.50	2.55	2.60	2.65	2.70	2.75	2.80	2.85	2.90	2.95	3.00	3.05	3.10		
0.19~0.23	2.30	2.35	2.40	2.45	2.50	2.55	2.60	2.65	2.70	2.75	2.80	2.85	2.90	2.95	3.00	3.05	3.10			
0.24~0.28	2.35	2.40	2.45	2.50	2.55	2.60	2.65	2.70	2.75	2.80	2.85	2.90	2.95	3.00	3.05	3.10				
0.29~0.33	2.40	2.45	2.50	2.55	2.60	2.65	2.70	2.75	2.80	2.85	2.90	2.95	3.00	3.05	3.10					
0.34~0.38	2.45	2.50	2.55	2.60	2.65	2.70	2.75	2.80	2.85	2.90	2.95	3.00	3.05	3.10						
0.39~0.43	2.50	2.55	2.60	2.65	2.70	2.75	2.80	2.85	2.90	2.95	3.00	3.05	3.10							
0.44~0.48	2.55	2.60	2.65	2.70	2.75	2.80	2.85	2.90	2.95	3.00	3.05	3.10								
0.49~0.53	2.60	2.65	2.70	2.75	2.80	2.85	2.90	2.95	3.00	3.05	3.10									
0.54~0.58	2.65	2.70	2.75	2.80	2.85	**2.90**	2.95	3.00	3.05	3.10										
0.59~0.63	2.70	2.75	2.80	2.85	2.90	2.95	3.00	3.05	3.10											
0.64~0.68	2.75	2.80	2.85	2.90	2.95	3.00	3.05	3.10												
0.69~0.73	2.80	2.85	2.90	2.95	3.00	3.05	3.10													
0.74~0.78	2.85	2.90	2.95	3.00	3.05	3.10														
0.79~0.83	2.90	2.95	3.00	3.05	3.10															
0.84~0.88	2.95	3.00	3.05	3.10																
0.89~0.93	3.00	3.05	3.10																	
0.94~0.98	3.05	3.10																		
0.99~1.03	3.10																			

I. Measure tappet clearance. "ENGINE COLD"
II. Measure present shim size.
III. Match clearance in vertical column with present shim size in horizontal column.

EXAMPLE

Tappet clearance is — 0.55 mm
Present shim size — 2.40 mm
Shim size to be used — 2.90 mm

Note: camlobe must be in position A or B when clearance is being measured.

Tappet shim selection chart

6 Carburettors: adjustment and synchronisation

In order that the engine maintains the best possible performance at all times, the carburettors must always remain correctly adjusted and synchronised. This check is essential and should be carried out at the specified intervals.

Synchronisation and adjustment of the carburettors requires the use of a set of four vacuum gauges or indicators, together with the appropriate adaptors which screw into the inlet tracks and to which are attached the vacuum take-off pipes. The adjustment of the carburettors is critical if smooth running and optimum fuel economy is to be expected and if damage to the engine due to incorrect mixture is to be avoided. Because of this and the prohibitive cost of the gauges required, it is recommended that the machine be returned to a Suzuki Service Agent, who will be able to carry out the work. If the vacuum gauges are available and some previous experience has been gained in their use, refer to Chapter 2, Sections 8, 9 and 10 for the prescribed adjustment procedure on the different models.

The idling speed of the engine may be altered by turning the remote adjuster (throttle stop screw) placed below the central instrument, of the bank of four carburettors. It is easily recognised by the large nylon head that it is fitted with. The correct idling speed is 1050 rpm $\pm$ 100 rpm. Always adjust the idling speed with the engine having been run so that it is operating at its normal working temperature.

For instruction on adjusting the twin throttle cables on the GN model refer to Chapter 2, Section 9.3, and apply the same procedure to adjust the single operating cable of the GT/GLT models.

7 Air filter cleaning

If the air filter element is permitted to become clogged with dust etc and regular servicing is ignored, the intake resistance will increase with a resultant decrease in engine performance and an accompanying rise in the fuel consumption.

The air filter case on the GS 850 models is fitted with a chromed cover at each end, both of which are retained by two screws.

Detach both covers and remove the single retaining screw from the left-hand side. The carrier and element may be pushed out. Detach the element from the carrier after removing the two remaining screws. The element is of the oil impregnated polyurethane type, and should be carefully washed out, dried and re-impregnated with engine oil. Full details of this procedure will be found in Chapter 2, Section 12.

8 Cleaning and checking the sparking plugs

Remove the sparking plugs and remove the carbon deposits using a wire brush. Clean the electrodes using fine emery paper or cloth, and then reset the gaps to 0.6 – 0.8 mm (0.024 – 0.031 in) using a feeler gauge. Whenever the plugs are removed, make a visual inspection of the condition of the plugs, especially the colour of the carbon deposits. See the sparking plug colour condition chart in Chapter 3. If the standard plug (NGK B8ES or Nippon Denso W24ES/W24ES-U) appears deficient; either appearing to be too hot (too soft a grade) or too cold (too hard a grade), then alternative plug types should be installed. If the plugs are apt to run too hot, that is, on examination the electrode is white in colour and has a blistered appearance, a colder range plug is necessary. A set of NGK B9ES or Nippon Denso W27ES plugs should be installed to correct this situation. Conversely, a black, sooty appearance to the plug, may indicate it is running too cold. In this instance a harder grade set of plugs; NGK B7ES or Nippon Denso W22ES, should be fitted to alleviate the problem.

Before replacing the plugs, smear the threads with a small amount of graphite grease to aid future removal.

9 Clutch adjustment

Accurate adjustment of the clutch is necessary to ensure efficient operation of the whole unit. With the type of clutch fitted to the GS 850 range, no adjustment is available on the clutch itself. Adjustment is possible at two points on the clutch cable, one at the operating lever, and a second at the actuating lever on the top of the right-hand side crankcase cover. To a certain extent the intervals at which the clutch should be adjusted will depend on the style of riding and the conditions in which the machine is used.

Pull back the rubber cover on the handlebar adjuster nut and lock nut. Slacken the locknut, and screw the adjuster fully inwards, to give maximum free play at the lever. Loosen the locknut on the actuating lever on the crankcase cover, and turn the chromed adjuster nut. This tensions the inner cable and reduces the play at the handlebar lever. When the play at the handlebar lever is between 2 – 3 mm (0.08 – 0.12 in), the adjustment is correct. Retighten the locknut at the lower cable and adjuster. If correct adjustment cannot be made by this method, further adjustment, at the handlebar, must be made. Screw out the adjuster until the free play is correct. Secure the locknut, and replace the rubber shroud.

10 Control cable lubrication

Use motor oil or an all purpose oil to lubricate the control cables. A good method for lubricating the cables is shown in the accompanying illustration, using a plasticine funnel. This method has a disadvantage in that the cables usually need removing from the machine. An hydraulic cable oiler which pressurises the lubricant overcomes this problem. Nylon lined cables should not be lubricated; in some cases the oil will cause the lining to swell leading to seizure.

Clean and check air filter element carefully and thoroughly

Use lower end cable adjuster to maintain correct clutch free play

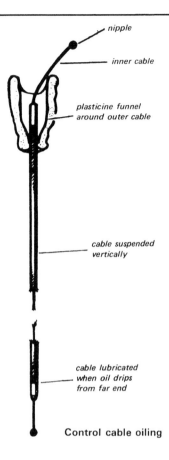

Control cable oiling

11 Cylinder head nuts and bolts torque check

This check is easiest to carry out after the tappet clearances have been checked and/or adjusted, whilst the petrol tank is removed. Using the tightening sequence given in Chapter 1, Section 52, tighten first the twelve 10 mm nuts to a torque setting of 3.7 kgf m (27.0 lbf ft), and then the three 6 mm bolts to a torque setting of 0.9 kgf m (6.5 lbf ft).

At this time, it is useful to check the tightness of the exhaust pipe flange bolts. They should be tightened to a torque setting of 0.9 – 1.2 kgf m (6.5 – 8.5 lbf ft).

12 Steering head bearings check

Make a general check on the condition of the steering head bearings, if necessary adjusting them as follows.

Place the machine on the centre stand so that the front wheel is clear of the ground. If necessary, apply a weight to the rear portion of the dualseat to prevent the machine tipping forwards. Grasp the front forks near the wheel spindle and push and pull firmly in a fore and aft direction. If play is evident between the upper and lower steering yokes and the head lug casting, the steering head bearings are in need of adjustment. Imprecise handling or a tendency for the front forks to judder may be caused by this fault.

To adjust the bearing, loosen the pinch bolt that passes through the rear of the upper yoke. Immediately below the upper yoke, on the steering stem, is the pegged adjuster nut. Using a C-spanner, tighten the adjuster nut a litle at a time until all play is taken up. **Do not** overtighten the nut. It is possible to place a pressure of several tons on the head bearings by over-tightening, even though the handlebars may seem to turn quite freely. Overtight bearings will cause the machine to roll at low speeds and give imprecise steering. Adjustment is correct if there is no play at the bearings and the handlebars swing to full lock either side, when the machine is on the centre stand, with the front wheel off the ground. Only a light tap on each end should cause the handlebars to swing.

Eight monthly or every 6000 miles (9600 km)

Again complete the checks listed under the previous routine maintenance interval headings. The following additional tasks are now necessary.

1 Secondary drive gear and the final drive gear oil change

Ensure the machine is supported on its centre stand. Remove the gear change operating lever, and the secondary drive gear casing cover from the left-hand side of the engine. The cover is retained by four crosshead screws. Place a suitable container below the rear, left-hand wall of the engine. Remove the oil filler cap bolt from the top of the casing, to aid drainage, and unscrew the drain plug from the lower part of the casing The hexagon headed drain plug is located directly below the plastic cap to the rear of the secondary drive bevel gear unit.

Drain the final drive gear case (at the rear wheel) in a similar manner. The drain plug is situated on the underside of the case. The two cases must be drained when the oil is warm, thereby ensuring that complete drainage is achieved. The final drive case particularly, takes an extended length of time to warm up, and therefore should be drained only after a long run.

Replenish the two cases with Hypoid gear oil (SAE 90), after checking that the drain plugs have been fitted and tightened. The approximate quantities are as follows:

Secondary drive casing	340-400 cc (11.5-13.5/12.0-14.1 US/Imp fl oz)
Final drive gear case	280-330 cc (9.5-11.2/9.9-11.6 US/Imp fl oz)

Check the level in the secondary drive casing by removing the oil level screw to the rear of the casing. When the oil starts to dribble out of the oil level screw orifice, the level is correct. Re-install the oil level screw when level is correct. Check the level in the final drive gear case visually. If the oil level is visible and roughly flush with the bottom of the filler cap bolt orifice, the level is correct.

2 Sparking plug renewal

Although in general sparking plugs will continue to function after this mileage, their efficiency will have been reduced. The correct standard plug type is NGK B8ES or Nippon Denso W24ES/W24ES-U. Before fitting, set the gaps to 0.6 – 0.8 mm (0.024 – 0.031 in).

3 Front fork air pressure check – US model only

Check, and if necessary, adjust the front fork air pressure (US models only). See Chapter 4, Section 6, paragraphs 6 – 8.

Yearly or every 12 000 miles (20 000 km)

Again carry out the tasks and checks listed under the previous routine maintenance interval headings. The following additional checks are now necessary.

1   Check the condition of the contact breaker point assemblies (GN models only), and renew them if necessary. See Chapter 3, Section 5.
2   Remove and clean the wheel bearings. Renew worn bearings. See Chapter 5, Sections 8 and 16.
3   In addition to the above operations, the various frame and engine fittings should be checked for tightness and lubrication applied where necessary.
4   In addition to all the foregoing, Suzuki recommend that at two yearly (or 24 000 mile) intervals the brake fluid hoses and the fuel lines be renewed.

Remove secondary drive gear units casing and remove drain plug

Replenish with required amount of Hypoid SAE 90 gear oil

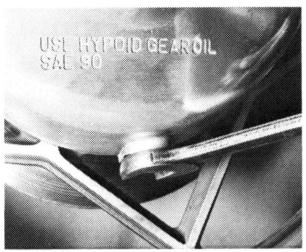

Remove final drive case drain bolt

Replenish case until oil level can be seen through filter aperture

General maintenance adjustments

It may have been noted that no reference has been made in any of the foregoing routine maintenance schedules to brake pad wear or inspection.

Brake pad wear depends largely on the conditions in which the machine is ridden and at what speed. It is difficult, therefore, to give precise inspection intervals, but it follows that pad wear should be checked more frequently on a hard ridden machine.

The condition of each pad can be checked easily whilst still in situ on the machine. The pads have a red groove around their outer periphery. This red groove can be seen from the front the caliper unit (front brakes) of the GN model, or from the top of the caliper after the inspection cap has been prised off (GT/GLT models). The rear pads may be checked in the same way as for the front pads of the GT/GLT models; by prising free the inspection cover.

Removal and replacement of the brake pads in both the front and rear calipers is a relatively simple operation, and may be accomplished without removing the respective wheel.

If wear has reduced either or both pads in one caliper down to the red line, the pads should be renewed as a pair. In practice, with the double front disc set-up used, if one set of pads requires renewal, it will be necessary to renew the other pair, too.

Full details of the renewal and replacement procedure of front and rear pads, are given in Chapter 5, Sections 4 and 11.

Quick glance
maintenance adjustments and capacities

Engine oil capacity
  Dry .... 3.8 lit (8.0/6.6 US/Imp pint)
  Oil change .... 2.8 lit (5.9/4.9 US/Imp pint)
  Oil and filter change .... 3.6 lit (7.6/6.4 US/Imp pint)

Front forks
  GS850 GN and GT (US models) .... 251 cc (8.49 US fl oz)
  GS850 GN and GT (UK models) .... 213 cc (7.50 Imp fl oz)
  GS850 GLT .... 302 cc (10.21 US fl oz)

Contact breaker gap (GN model) .... 0.3 – 0.4 mm (0.012 – 0.016 in)

Sparking plug gap .... 0.6 – 0.8 mm (0.023 – 0.031 in)

Valve clearances (cold)
  Inlet .... 0.03 – 0.08 mm (0.001 – 0.003 in)
  Exhaust .... 0.03 – 0.08 mm (0.001 – 0.003 in)

Tyre pressures

	Solo	Pillion
Front	25 psi (1.75 kg/cm^2)	25 psi (1.75 kg/cm^2)
Rear	28 psi (2.00 kg/cm^2)	32 psi (2.25 kg/cm^2)

For continous high-speed riding, the pressures should be increased to:

	Solo	Pillion
Front	28 psi (2.00 kg/cm^2)	28 psi (2.00 kg/cm^2)
Rear	32 psi (2.25 kg/cm^2)	40 psi (2.80 kg/cm^2)

Recommended lubricants

Component	Lubricant
Fuel grade	Unleaded or low-lead, minimum octane rating 90 RON/RM
Engine and gearbox	SAE 10W/40 engine oil, API class SE or SF
Secondary and final drives	Hypoid gear oil SAE 90, API class GL-5 (If ambient temperature is below 0°C (32°F) use SAE 80 gear oil)
Front forks	SAE 10W/20 fork oil
Wheel bearings	Lithium base, high melting point grease
Disc brakes	Hydraulic brake fluid conforming to DOT 3 or 4 (USA) or SAE J1703 (UK)

English/American terminology

Because this book has been written in England, British English component names, phrases and spellings have been used throughout. American English usage is quite often different and whereas normally no confusion should occur, a list of equivalent terminology is given below.

English	American	English	American
Air filter	Air cleaner	Number plate	License plate
Alignment (headlamp)	Aim	Output or layshaft	Countershaft
Allen screw/key	Socket screw/wrench	Panniers	Side cases
Anticlockwise	Counterclockwise	Paraffin	Kerosene
Bottom/top gear	Low/high gear	Petrol	Gasoline
Bottom/top yoke	Bottom/top triple clamp	Petrol/fuel tank	Gas tank
Bush	Bushing	Pinking	Pinging
Carburettor	Carburetor	Rear suspension unit	Rear shock absorber
Catch	Latch	Rocker cover	Valve cover
Circlip	Snap ring	Selector	Shifter
Clutch drum	Clutch housing	Self-locking pliers	Vise-grips
Dip switch	Dimmer switch	Side or parking lamp	Parking or auxiliary light
Disulphide	Disulfide	Side or prop stand	Kick stand
Dynamo	DC generator	Silencer	Muffler
Earth	Ground	Spanner	Wrench
End float	End play	Split pin	Cotter pin
Engineer's blue	Machinist's dye	Stanchion	Tube
Exhaust pipe	Header	Sulphuric	Sulfuric
Fault diagnosis	Trouble shooting	Sump	Oil pan
Float chamber	Float bowl	Swinging arm	Swingarm
Footrest	Footpeg	Tab washer	Lock washer
Fuel/petrol tap	Petcock	Top box	Trunk
Gaiter	Boot	Torch	Flashlight
Gearbox	Transmission	Two/four stroke	Two/four cycle
Gearchange	Shift	Tyre	Tire
Gudgeon pin	Wrist/piston pin	Valve collar	Valve retainer
Indicator	Turn signal	Valve collets	Valve cotters
Inlet	Intake	Vice	Vise
Input shaft or mainshaft	Mainshaft	Wheel spindle	Axle
Kickstart	Kickstarter	White spirit	Stoddard solvent
Lower leg	Slider	Windscreen	Windshield
Mudguard	Fender		

Working conditions and tools

When a major overhaul is contemplated, it is important that a clean, well-lit working space is available, equipped with a workbench and vice, and with space for laying out or storing the dismantled assemblies in an orderly manner where they are unlikely to be disturbed. The use of a good workshop will give the satisfaction of work done in comfort and without haste, where there is little chance of the machine being dismantled and reassembled in anything other than clean surroundings. Unfortunately, these ideal working conditions are not always practicable and under these latter circumstances when improvisation is called for, extra care and time will be needed.

The other essential requirement is a comprehensive set of good quality tools. Quality is of prime importance since cheap tools will prove expensive in the long run if they slip or break and damage the components to which they are applied. A good quality tool will last a long time, and more than justify the cost. The basis of any tool kit is a set of open-ended spanners, which can be used on almost any part of the machine to which there is reasonable access. A set of ring spanners makes a useful addition, since they can be used on nuts that are very tight or where access is restricted. Where the cost has to be kept within reasonable bounds, a compromise can be effected with a set of combination spanners – open-ended at one end and having a ring of the same size on the other hand. Socket spanners may also be considered a good investment, a basic $\frac{3}{8}$ in or $\frac{1}{2}$ in drive kit comprising a ratchet handle and a small number of socket heads, if money is limited. Additional sockets can be purchased, as and when they are required. Provided they are slim in profile, sockets will reach nuts or bolts that are deeply recessed. When purchasing spanners of any kind, make sure the correct size standard is purchased. Almost all machines manufactured outside the UK and the USA have metric nuts and bolts, whilst those produced in Britain have BSF or BSW sizes. The standard used in the USA is AF, which is also found on some of the later British machines. Other tools that should be included in the kit are a range of crosshead screwdrivers, a pair of pliers and a hammer.

When considering the purchase of tools, it should be remembered that by carrying out the work oneself, a large proportion of the normal repair cost, made up by labour charges, will be saved. The economy made on even a minor overhaul will go a long way towards the improvement of a tool kit.

In addition to the basic tool kit, certain additional tools can prove invaluable when they are close to hand, to help speed up a multitude of repetitive jobs. For example, an impact screwdriver will ease the removal of screws that have been tightened by a similar tool, during assembly, without a risk of damaging the screw heads. And, of course, it can be used again to retighten the screws, to ensure an oil or airtight seal results. Circlip pliers have their uses too, since gear pinions, shafts and similar components are frequently retained by circlips that are not too easily displaced by a screwdriver. There are two types of circlip pliers, one for internal and one for external circlips. They may also have straight or right-angled jaws.

One of the most useful of all tools is the torque wrench, a form of spanner that can be adjusted to slip when a measured amount of force is applied to any bolt· or nut. Torque wrench settings are given in almost every modern workshop or service manual, where the extent is given to which a complex component, such as a cylinder head, can be tightened without fear of distortion or leakage. The tightening of bearing caps is yet another example. Overtightening will stretch or even break bolts, necessitating extra work to extract the broken portions.

As may be expected, the more sophisticated the machine, the greater is the number of tools likely to be required if it is to be kept in first class condition by the home mechanic. Unfortunately there are certain jobs which cannot be accomplished successfully without the correct equipment and although there is invariably a specialist who will undertake the work for a fee, the home mechanic will have to dig more deeply in his pocket for the purchase of similar equipment if he does not wish to employ the services of others. Here a word of caution is necessary, since some of these jobs are best left to the expert. Although an electrical multimeter of the AVO type will prove helpful in tracing electrical faults, in inexperienced hands it may irrevocably damage some of the electrical components if a test current is passed through them in the wrong direction. This can apply to the synchronisation of twin or multiple carburettors too, where a certain amount of expertise is needed when setting them up with vacuum gauges. These are, however, exceptions. Some instruments, such as a strobe lamp, are virtually essential when checking the timing of a machine powered by CDI ignition system. In short, do not purchase any of these special items unless you have the experience to use them correctly.

Although this manual shows how components can be removed and replaced without the use of special service tools (unless absolutely essential), it is worthwhile giving consideration to the purchase of the more commonly used tools if the machine is regarded as a long term purchase. Whilst the alternative methods suggested will remove and replace parts without risk of damage, the use of the special tools recommended and sold by the manufacturer will invariably save time.

Chapter 1 Engine, clutch and gearbox

Refer to Chapter 7 for information relating to the 1981 to 1988 models

Contents

Specifications

Engine

Type	Four cylinder, double overhead camshaft, air-cooled, four-stroke
Bore	69.0 m (2.717 in)
Stroke	56.4 mm (2.220 in)
Capacity	843 cc (51.4 cu in)
Compression ratio	8.8:1
Maximum bhp	77 @ 8500 rpm

Cylinder head
    Max. cylinder head warpage .... 0.2 mm (0.008 in)

Valves and springs
    Valve stem outside diameter:
        Inlet .... 6.960 – 6.975 mm (0.2740 – 0.2746 in)
        Exhaust .... 6.945 – 6.960 mm (0.2734 – 0.2740 in)
    Valve head diameter:
        Inlet .... 36.0 mm (1.415 in)
        Exhaust .... 30.0 mm (1.185 in)
    Valve stem run-out (max) .... 0.05 mm (0.0020 in)
    Valve guide internal diameter:
        Inlet/exhaust .... 7.000 – 7.015 mm (0.2756 – 0.2762 in)
    Valve stem to guide clearance:
        Inlet .... 0.025 – 0.055 mm (0.0010 – 0.0022 in)
        Service limit .... 0.090 mm (0.0035 in) – GN (UK), 0.350 mm (0.0138 in) – all other models
        Exhaust .... 0.040 – 0.070 mm (0.0016 – 0.0028 in)
        Service limit .... 0.100 mm (0.0039 in) – GN (UK), 0.350 mm (0.0138 in) – all other models
    Valve spring free length:
        Inner .... 35.3 – 37.0 mm (1.39 – 1.46 in)
        Service limit .... 33.9 mm (1.33 in)
        Outer .... 43.0 – 43.25 mm (1.69 – 1.70 in)
        Service limit .... 41.3 mm (1.63 in)

Camshafts
    Overall lobe height:
        Inlet .... 36.320 – 36.360 mm (1.4299 – 1.4315 in)
        Service limit .... 36.020 mm (1.4181 in)
        Exhaust .... 35.770 – 35.810 mm (1.4083 – 1.4098 in)
        Service limit .... 35.470 mm (1.3965 in)
    Journal outside diameter .... 21.960 – 21.975 mm (0.8646 – 0.8652 in)
    Journal bearing inside diameter:
        GS850 GN (UK) .... 22.000 – 22.013 mm (0.8661 – 0.8667 in)
        All other models .... 22.012 – 22.025 mm (0.8666 – 0.8671 in)
    Journal to bearing clearance:

	Standard	Service limit
GS850 GN (UK)	0.025 – 0.053 mm (0.0010 – 0.0021 in)	0.150 mm (0.0059 in)
All other models	0.037 – 0.065 mm (0.0015 – 0.0026 in)	0.150 mm (0.0059 in)

    Camshaft run-out (max) .... 0.1 mm (0.004 in)

Piston and rings
    Piston diameter .... 68.945 – 68.960 mm (2.7144 – 2.7150 in)
    Service limit .... 68.880mm (2.7118 in)
    Gudgeon pin bore internal diameter in piston .... 16.002 – 16.008 mm (0.6300 – 0.6302 in)
    Service limit .... 16.030 mm (0.6311 in)
    Gudgeon pin outside diameter .... 15.995 – 16.000 mm (0.6297 – 0.6300 in)
    Service limit .... 15.980 mm (0.6291 in)
    Gudgeon pin/pin bore clearance .... 0.002 – 0.013 mm (0.0001 – 0.0005 in)
    Service limit .... 0.12 mm (0.0047 in)
    Piston ring groove width:
        Top and 2nd ring .... 1.21 – 1.23 mm (0.047 – 0.048 in)
        Oil control ring .... 2.51 – 2.53 mm (0.099 – 0.100 in)
    Piston ring free end gap:
        Top ring .... 9.0 mm (0.35 in)
        Service limit .... 7.2 mm (0.28 in)
        2nd ring .... 9.5 mm (0.37 in)
        Service limit .... 7.6 mm (0.30 in)
    Piston ring installed end gap:
        Top and 2nd ring .... 0.1 – 0.3 mm (0.004 – 0.012 in)
        Service limit .... 0.7 mm (0.03 in)
    Piston ring thickness:
        Top ring .... 1.175 – 1.190 mm (0.0463 – 0.0469 in)
        2nd ring .... 1.170 – 1.190 mm (0.0461 – 0.0469 in)
    Piston ring side clearance:
        Top ring .... 0.020 – 0.055 mm (0.0008 – 0.0022 in)
        Service limit .... 0.18 mm (0.0071 in)
        2nd ring .... 0.020 – 0.060 mm (0.0008 – 0.0024 in)
        Service limit .... 0.15 mm (0.0059 in)

Cylinder barrels
    Bore diameter .... 69.000 – 69.015 (2.7165 – 2.7171 in)
    Service limit .... 69.080 mm (2.7197 in)
    Cylinder/piston clearance .... 0.050 – 0.060 mm (0.0020 – 0.0024 in)
    Service limit .... 0.120 mm (0.0047 in)

Compression pressure	9 – 12 kg/cm² (128 – 171 psi)
Service limit	7 kg/cm² (100 psi)
Pressure difference between cylinders (max)	2 kg/cm² (28 psi)

Crankshaft and connecting rods

Crankshaft run-out (max)	0.05 mm (0.002 in)
Connecting rod deflection (max)	3.0 mm (0.12 in)
Connecting rod axial float	0.10 – 0.55 mm (0.004 – 0.026 in)
Service limit	1.0 mm (0.039 in)
Big-end bearing radial wear (max)	0.08 mm (0.031 in)
Connecting rod small-end bore	16.006 – 16.014 mm (0.6302 – 0.6305 in)
Service limit	16.040 mm (0.6315 in)

Clutch

Type	Wet, multi-plate
No of plates:	
Plain	8
Friction	8
Plain plate thickness	2.0 ± 0.06 mm (0.08 ± 0.002 in)
Plain plate warpage (max)	0.1 mm (0.004 in)
Friction plate thickness	2.7 – 2.9 mm (0.106 – 0.114 in)
Service limit	2.4 mm (0.094 in)
Friction plate warpage (max)	0.2 mm (0.008 in)
No of springs	6
Spring free length	40.4 mm (1.59 in)
Service limit	38.5 mm (1.52 in)

Gearbox

Type	5-speed, constant mesh
Gear ratios (no of teeth):	
1st gear	2.500 : 1 (35/14)
2nd gear	1.777 : 1 (32/18)
3rd gear	1.380 : 1 (29/21)
4th gear	1.125 : 1 (27/24)
5th gear	0.961 : 1 (25/26)
Primary reduction	1.775 : 1 (87/49)
Secondary reduction	1.062 : 1 (17/16)
Final drive ratio	3.090 : 1 (34/11)

Main torque wrench settings

Cylinder head:	
6 mm bolts	0.9 kgf m (6.5 lbf ft)
10 mm nuts	3.7 kgf m (27.0 lbf ft)
Cylinder head cover bolts	0.9 kgf m (6.5 lbf ft)
Engine mounting bolts:	
8 mm bolts	2.5 kgf m (18.0 lbf ft)
10 mm bolts	3.5 kgf m (25.5 lbf ft)
Crankcase bolts:	
6 mm bolts	1.0 kgf m (7.5 lbf ft)
8 mm bolts	2.0 kgf m (14.5 lbf ft)
Camshaft cap bolts	1.0 kgf m (7.5 lbf ft)
Clutch spring bolts	1.1 – 1.3 kgf m (8.0 – 9.5 lbf ft)
Clutch centre nut	5.0 – 7.0 kgf m (36.0 – 50.5 lbf ft)
Gear change operating lever bolt	1.3 – 2.3 kgf m (9.5 – 16.5 lbf ft)
Alternator rotor centre bolt	9.0 – 10.0 kgf m (65.0 – 72.5 lbf ft)
Secondary drive gear nut	12.0 – 15.0 kgf m (87.0 – 108.5 lbf ft)
Secondary driven gear nut	9.0 – 11.0 kgf m (65.0 – 79.5 lbf ft)
Secondary drive housing bolts	2.0 – 2.6 kgf m (14.5 – 19.0 lbf ft)
Secondary driven housing bolts	2.0 – 2.6 kgf m (14.5 – 19.0 lbf ft)

1 General description

The engine unit fitted to the Suzuki GS850 model is of the four-cylinder, air cooled, in-line type, fitted transversely across the frame. The valves are operated by double overhead camshafts driven from the crankshaft by a centre chain. The two camshafts are located in the cylinder head casting, and the camshaft chain drive operates through a cast-in tunnel between the four cylinders. Adjustment of the chain is effected by a chain tensioner, fitted to the rear of the cylinder block. The chain tensioner is of the automatic, self-adjusting type maintaining correct tension on the chain and compensating for chain wear, after the initial adjustment has been made during assembly.

The engine/gear unit is of aluminium alloy construction, with the crankcases dividing horizontally.

Lubrication is of the pressure-feed, wet-sump type. The system incorporates a gear driven oil pump, an oil filter, a safety by-pass valve, and an oil pressure switch.

Oil vapours created in the crankcase are vented through an

oil breather to the air cleaner case where they are recirculated into the crankcase, providing an oil tight system.

An Eaton trochoid oil pump is fitted, driven by a gear pinion to the rear of the clutch outer drum. Oil is picked up from the sump via a chamber integral with the upper crankcase half, the mouth of which is closed by a detachable wire mesh screen. The screen protects the oil pump from any larger impurities which may have contaminated the oil. The oil pump forces the oil, under pressure, through a full flow paper-element oil filter which is housed within a chamber at the front of the crankcase. In the event of the filter becoming blocked, a by-pass valve is included which prevents cessation of the oil flow by opening at a preset pressure.

After passing through the oil filter, the oil flow is separated into three branch systems. The main system feeds the crankshaft main bearings and the big-end bearings, and the two remaining systems supply the camshafts and valves and the gearbox shafts and pinions. Returning oil from the engine and gearbox components falls under gravity to the sump, where it is picked up once more by the oil pump and the cycle is repeated.

Engine power from the crankshaft is transmitted directly to the gearbox mainshaft by a large helical cut pinion attached to the rear of the clutch outer drum. Coil spring shock absorbers are mounted in the rear of the clutch drum, to damp out snatch loads in the transmission.

The five-speed, constant-mesh gearbox is placed to the rear of the crankshaft in the normal manner. The mainshaft is of unusual construction, being in effect a shaft within a shaft, the two being interconnected by a spring loaded cam face shock absorber. Shock loads in the transmission are damped out by the differential movement in the two concentric shafts, allowed by the cam faces riding up each other. The layshaft, positioned to the rear of the mainshaft, transmits power to the secondary drive bevel gear. This in turn engages with the secondary driven bevel gear which turns the drive line through 90° and so enables the final drive shaft to be interconnected. The drive shaft itself runs in the left-hand side of the swinging arm and makes a further 90° turn at the rear hub.

2 Operations with the engine/gearbox unit in the frame

1 It is not necessary to remove the engine from the frame to carry out certain operation; in fact it can be an advantage. Tasks that can be carried out with the engine in situ are as follows:
a) Removal of cylinder head, cylinder block and pistons.
b) Removal of the clutch/primary drive gear.
c) Removal of the alternator and starter motor.
d) Removal of carburettors.
e) Removal of the gearchange external selector mechanism.
f) Removal of the oil pump and oil filter.
g) Removal of the secondary drive bevel gear assembly.
h) Removal of the secondary driven bevel gear assembly.
2 When several tasks have to be undertaken simultaneously, it will probably be advantageous to remove the complete engine unit from the frame, an operation that should take about an hour and a half. This gives the advantage of much better access and more working space.

3 Operations with the engine/gearbox unit removed from the frame

1) Removal of the crankshaft assembly, complete with main bearings and connecting rod assemblies.
b) Removal of the gearbox components including the gearchange internal selection mechanism.
c) Removal of the kickstart shaft and engagement mechanism, (GS 850 GN model only).

4 Method of engine/gearbox removal

As described previously, the engine and gearbox are a built in unit and it is necessary to remove the unit complete in order to gain access to either unit. Separation of the crankcase is achieved after the engine unit has been removed from the frame and refitting cannot take place until the engine/gear unit is assembled completely. Access to the gearbox is not available until the engine has been dismantled and vice-versa in the case of attention to the bottom end of the engine. Fortunately, the task is made easy by arranging the crankcase to separate horizontally.

5 Removing the engine/gearbox unit

1 Place the machine on the centre stand, so that it is standing firmly on level ground. Place a receptacle that will hold at least a gallon under the crankcase and remove the drain plug so that the oil will drain off. It is preferable to do this whilst the engine is warm, so that the oil will drain more readily.
2 Transfer the container so that it rests below the oil filter chamber at the front of the engine. Remove the three domed nuts and the washers and detach the cover. The cover is spring loaded by the filter element retainer spring and so should be released in a controlled manner. Lift out the spring and the element.
3 If the previous container is not full, move it to a position below the secondary drive/driven bevel gear assemblies on the left-hand side of the machine. Remove the gearchange lever from the splined operating shaft. The lever is retained by a pinch bolt which must be unscrewed fully before the lever can be displaced. Remove the screws which secure the secondary drive unit cover, and detach the cover. Remove the drain plug so that the gear oil will drain off. To aid complete drainage, remove the filler cap from the top of the chamber.
4 Unlock and raise or remove the dualseat, and lift out the tool kit and its tool tray/battery cover. Disconnect both of the battery leads from the terminals. Isolating the electrical system in this way will prevent accidental shorting of wires which are subsequently disconnected. If it is expected that the machine is to be unused for a protracted length of time the battery should be removed at this stage and given a refresher charge from an independent source at approximately monthly intervals. The battery may be lifted from the carrier after displacing the retainer strap and pulling the small breather hose from the union at the side of the battery.
5 Place the petrol tap lever in the 'On' or 'Reserve' position on the GS 850 GN and GT models, or ensure the tap is in its normal operating mode on the GLT model, and disconnect from the tap unions the petrol feed pipe and the small vacuum pipe which controls the tap diaphragm. The feed pipe is secured by a spring clip, the ears of which should be squeezed together to release the grip on the pipe. Remove the single bolt which passes through the lug at the rear of the petrol tank and into the frame. The tank is supported at the front by two steel cups which locate with a rubber buffer each side of the frame top tube. Before removing the petrol tank the fuel gauge sensor leads must be disconnected. Raise the rear of the tank to expose the two fuel gauge leads at the lower left-hand side of the petrol tank. Disconnect the leads at their snap connectors. Ease the tank rearwards until the cups clear the buffers and then lift it away. Drainage of the tank is not strictly necessary although to do so will reduce the overall weight and hence facilitate removal. A full tank will weigh in the region of 35 lb.
6 On the GS 850 GLT model, detach the two plastic control cable shrouds; each is secured by two screws, which are fitted to each side of the frame top tube as it leaves the steering stem.
7 Detach the left-hand and right-hand frame side covers. Each cover is a push fit on the frame, being secured by a projection at the lower edge which is a push fit in a rubber grommet, and at the upper edge on two rubber covered hook tabs projecting from the frame.
8 Pull off the engine breather hose from the breather cover union on top of the cam cover. The hose is secured by a spring clip, the ears of which must be squeezed together to release the grip on the hose.

9   Remove the single air cleaner case retaining bolt from each side of the machine. The case retaining bolts pass through a small lug projecting from the top frame tubes. Loosen the screw clips which secure the carburettors to the inner stubs, and those securing the air filter hoses to the carburettor mouths. The air filter case is now free to be removed. Before doing so, however, disconnect all the wiring which passes above and behind the air filter case at the individual snap connectors and the multiple block connector. Detach the breather and vent pipes from the carburettors which pass through the clip on the rear of the air filter case. Ease the air filter case slightly rearwards, so that the air filter hoses leave the carburettors. The air filter case can now be manoeuvred free from the confines of the frame, being withdrawn to the right-hand side of the machine.

10 On the GS 850 GN model, disconnect both throttle cables from the operating pulley at the carburettors. Each may be detached in a similar manner. Loosen the upper and lower locknuts on the cable adjuster screw and displace the adjuster and outer cable from the abutment bracket. Rotate the pulley until the nipple can be pushed out. Note that when detaching the throttle cables, the strong return spring tension will have to be overcome; ensure the spring is not released sharply whilst there are fingers in the vicinity. On the GS 850 GT/GLT models, there is only one throttle cable; this can be detached in a manner similar to that just described. On all models, the choke operating cable should also be displaced now. Free the lower end nipple from its operating arm, then slacken off the cable adjuster and release the cable. Pull the carburettors rearwards to free them from the inlet stubs, and remove then as a complete unit.

11 Remove the chromed cover which encloses the starter motor, and is held by two bolts. Pull back the protective rubber boot from the terminal projecting from the starter motor body, and detach the heavy cable from the terminal. Pull the lead from the top of the oil pressure warning switch. The lead is a push fit. Pull both leads from where they are routed and lodge them out of harm's way, towards the rear of the machine. Follow the leads from the alternator up to the rubber shroud to the electrical component mounting plate on the left-hand side of the machine. Peel back the shroud and disconnect the three wires – white/green tracer, yellow, and white/blue tracer – at their individual snap connectors. On the GS 850 GN model, the two leads from the contact breakers are similarly protected by a rubber sleeve, forward of the left-hand rear frame down tube. Disconnect both the black wire and white wire. If the block connector for the digital gear indicator unit was not separated earlier, follow the lead up from the pick-up on the left-hand wall of the gearbox, located inset and below the secondary drive assembly, and disconnect the block connector. Similarly, the single lead (blue) for the neutral indicator switch can be traced from the same source, bound together with the gear indicator leads, and separated at its individual snap connector. On the GS 850 GT/GLT models, separate the green wire and the blue wire at the connector at the electric ignition 'control box' on the electrical component mounting plate. These two wires emanate from the signal generating unit on the right-hand side of the engine in the space occupied by the contact breaker points on the GN model. All wires leading from the engine should be arranged at the rear of the engine so that they do not become snagged on engine removal.

12 Pull the plug caps from the sparking plugs and secure the HT leads to the upper frame tubes. Each lead is numbered, to aid correct replacement. Note that there are retaining clips on each of the plug leads; these should be bent out of the way to enable the leads to be freed and secured above the engine.

13 Moving to the rear of the engine, slacken the small screw which secures the screw clip on the drive shaft rubber boot. Carefully, prise the rubber boot rearwards to expose the flanged coupling, the drive shaft and the flange on the rear of the secondary driven bevel gear unit. An assistant will be helpful at this juncture. Instruct your assistant to operate the rear brake pedal whilst the four flange bolts are being slackened. Slacken evenly the bolts, using a ring spanner initially, followed by an ordinary open-end spanner. The ring spanner cannot be used to carry out complete removal of the bolts, because the bolt head traps the spanner against the flange. As each bolt is loosened. rotate the flange as required by turning the rear wheel, ensuring of course, that your faithful assistant is not still applying the rear brake. With all the bolts slackened, remove them.

14 Remove the rear brake pedal from the splined shaft, after unscrewing fully the pinch bolt. Prise the end of the pedal return spring from the anchor peg, with the aid of a stout screwdriver blade, and remove the pedal and spring together. Remove the two bolts that retain the right-hand side footrest, and remove the footrest. On the GS 850 GN model, slacken and remove the kickstart lever pinch bolt, and slide the lever off its splined shaft.

15 Slacken the bolts holding the four exhaust pipe to silencer joint clamps. On the GS850 GT/GLT models, which are fitted with a small silencer chamber interconnecting the exhaust pipes of cylinders No 2 and 3 below the sump pan, slacken all four of the clamps on the No 2 and 3 exhaust pipes. This will enable the silencer chamber to be detached separately. Slide each clamp back up the pipe to clear the joint. Note the use of fibre packing pieces at each joint. On the GS 850 GN model, the outermost (Nos 1 and 4) exhaust pipes are separate and may be removed individually. On the GS 850 GT/GLT models, the inner pair (Nos 2 and 3) of exhaust pipes are separate from the rest of the system. Commencing with either outer exhaust pipe (GN model) or either inner exhaust pipe (GT/GLT models), slacken evenly and then remove the two bolts securing each finned exhaust port flange. When removing these exhaust port bolts, care must be taken not to damage the internal threads. With their exposed position at the front of the engine, and with the effects of continual heating and cooling, the bolts tend to stick in the relatively soft alloy of the cylinder head. For this reason, some patience may be needed to withdraw these bolts without destroying their heads. If it is found that the bolts are indeed stuck in position, use copious amounts of penetrating oil and ease the bolts up and down the threads until they become free and can be removed. Do not be tempted simply to use brute force. Slide the flange down the pipe and then ease the pipe simultaneously from the exhaust port and silencer or exhaust port and silencer chamber. On the GT/GLT models, remove the silencer chamber by pulling it free from the silencers. Removal of the silencers is not strictly necessary to allow the exhaust system to be separated from the engine. It may, however, be easier to detach them, before lifting the system away from the machine, as the weight is considerable. Their removal will also avoid unnecessary damage to what are expensive items. Each silencer is retained by two bolts passing through lugs on the bottom of the unit and into the rear of the pillion footrest sub frame. When manoeuvring the remainder of the exhaust system clear of the machine it will prove easier if the side stand is lowered to clear the inner pipe/silencer connection or outer pipe/silencer connection.

16 From the right-hand side of the machine, remove the clutch cable and its release arm from the top of the clutch cover. Displace the rubber shroud on the handlebar control lever and screw the knurled adjuster inwards fully. It is not necessary to disconnect the cable from the handlebar lever unless examination is required. Pull the rubber boot back on the operating arm mechanism and loosen the locknut to allow the adjusting nut to be slackened fully. Slacken and remove the pinch bolt on the release arm and slide the arms upwards off the splined shaft.

17 To improve the clearance between the top of the engine and the upper frame tubes, and so aid engine removal, the two horns should be detached. Disconnect the two snap connectors on each horn and rest the leads on the frame tubes above the engine. Remove the single bolt which re⁺ains each horn to the shared mounting bracket, and remove the horns.

18 For the same reason, that of limited clearance between the engine and frame tubes, the breather cover attached to the top of the cambox should be detached. Remove the four bolts, noting the positioning of the HT lead clips on the two forwardmost bolts, noting the positioning of the HT lead clips on the two forwardmost bolts, and remove the cover. If the

cover is reluctant to lift clear, the cover gasket has probably become stuck to the cambox top. Carefully insert a flat bladed screwdriver around the periphery of the breather cover until the joint is broken and lift the cover away. Patience should be exercised when separating the cover to avoid damage to the mating surfaces. Detach the tachometer cable from the front of the cylinder head; the cable is held by a screwed ring.

19 The engine/gearbox unit of the GS 850 models is an extremely large, heavy and cumbersome component, and as such is not easily manhandled. It is recommended, therefore, that at least three people are present when lifting the engine from place. This alleviates the risk of both mechanical damage to the engine and personal injury to the operator. Before attempting to remove any of the engine mounting bolts, the aquisition of a jack will be very beneficial. By positioning the jack below the sump, much of the weight can be taken when the mounting bolts are removed. Although the use of a jack is not essential, it is strongly recommended that a jack be employed, for the safety of all concerned.

20 Remove first the front three mounting bolts and their mounting bracket from the right-hand side with the nut on the left-hand end of the long through bolt removed. Similarly, remove the front two mounting bolts – two short bolts – and their bracket from the left-hand side. Then remove the two upper rear mounting bolts, the nut and the bracket from the left-hand side, and the three upper rear mounting bolts and bracket on the right-hand side. Remove the single lower rear bolt on the right-hand side, and withdraw the long through bolt out to the right-hand side. Lastly, remove the three bolts and their bracket fitted below the right-hand side of the engine and the single bolt in the similar position on the left-hand side. Note that the centre bolt of the three on the right-hand side and the single left-hand side bolt of these centrally positioned mounting bolts, are fitted with triangular threaded plate nuts in place of normal nuts, which are restrained in recesses in the crankcase casting. After removal of all the bolts, the engine will have settled onto the jack, or be resting in the frame 'cradle'. The operator and his assistants should now have time to find a suitable handhold. Lift the engine up gradually, ensuring the secondary drive output flange does not damage the driveshaft coupling rubber boot and lift the engine out of the frame from the right-hand side.

5.9a Remove the air filter case retaining bolts

5.9b Displace the air filter case before carburettor removal

5.10a Remove both operating cables (GN model) from the pulley and ...

5.10b ... also displace the choke operating cable before ...

5.10c ... removing the carburettors as a complete unit

5.11 Remove the starter motor cover and detach lead from oil pressure warning switch (arrowed)

5.13 Remove four bolts to separate output flange from drive shaft

5.14a Remove brake pedal and return spring, and two bolts and footrest

5.14b On GN model only, remove the pinch bolt and kickstart lever

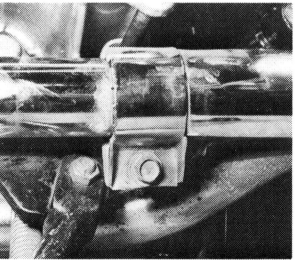

5.15a Slacken and pull back exhaust clamps to clear packing

5.15b Carefully remove flange bolts and detach exhaust pipes

5.15c Each silencer is retained by a bolt at the rear

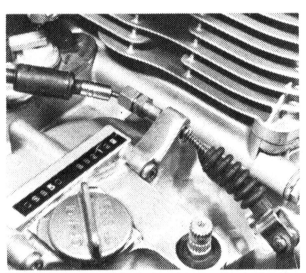

5.16 With adjustment fully slackened, remove the clutch cable release arm from its splined shaft

5.18 Remove breather cover to increase engine to frame clearance

5.20a Remove first the front three mounting bolts and bracket ...

5.20b ... and then the two left-hand short bolts and bracket

5.20c Remove this nut and two bolts from left-hand side ...

5.20d ... followed by three bolts on right-hand side

5.20e Remove single long bolt at lower rear of engine

5.20f Central bolt screws into plate nut – remove all three bolts and one similar on left-hand side

5.20g Lift out engine to right-hand side

6 Dismantling the engine and gearbox: general

1   Before commencing work on the engine unit, the external surfaces should be cleaned thoroughly. A motor cycle engine has very little protection from road grit and other foreign matter, which will find its way into the dismantled engine if this simple precaution is not taken. One of the proprietary cleaning compounds, such as Gunk or Jizer can be used to good effect, particularly if the compound is permitted to work into the film of oil and grease before it is washed away. Special care is necessary when washing down to prevent water from entering the now exposed parts of the engine unit.
2   Never use undue force to remove any stubborn part unless specific mention is made of this requirement.There is invariably good reason why a part is difficult to remove, often because the dismantling operation has been tackled in the wrong sequence. Dismantling will be made easier if a simple engine stand is constructed to correspond with the engine mounting points. This arrangement will permit the complete unit to be clamped rigidly to the workbench, leaving both hands free.

7 Dismantling the engine unit: removing the camshaft cover and the camshafts

1 Remove the four end caps from the camshaft cover; they may well require the use of an impact driver to free them successfully. Unscrew the 16 bolts which hold the camshaft cover in position, noting the positioning of HT lead clips in two of the bolts. If the cover is stuck firmly to the gasket, a soft-headed mallet may be used to break the seal. Use the mallet judiciously, striking only those parts of the cover which are well strengthened by lugs.

2 Unscrew the sparking plugs. On the GS 850 GN model, remove the contact breaker points cover from the engine right-hand casing. On the GS 850 GT/GLT models, remove the similar cover over the ignition signal generating unit. Using the correct sized spanner applied to the large hexagon fitted to the end of the contact breaker cam (GN model) or timing unit cam (GT/GLT models), rotate the engine forwards until the piston in the left-hand cylinder (No 1 cylinder) is at top dead centre (TDC). TDC may be found by viewing the timing marks on the Automatic Timing Unit (ATU). These marks may be viewed through the circular inspection aperture in the contact breaker stator plate (GN model) or the ignition signal generating unit mounting plate (GT/GLT models). Turn the engine until the T mark, which is to the left of the F1-4 mark, is in alignment with the index pointer in the casing.

3 Removal of the automatic cam chain tensioner must be carried out in a special sequence. Commence by loosening the locknut securing the grubscrew in the left-hand side of the tensioner body. Turn the screw inwards so that it tightens against the plunger within the body. The tensioner can now be removed without the spring loaded plunger being displaced. Remove the three mounting bolts to free the tensioner.

4 Detach the jockey sprocket assembly from between the two camshaft sprockets by removing the four mounting bolts. Mark the sprocket carrier so that it may be refitted in the same position. Remove its single mounting bolt and withdraw the tachometer drive assembly.

5 Having removed the tensioner and jockey sprocket, the camshafts may be removed individually, without separating the cam chain. It will be noted that no matter in what position the engine is placed, at least one cam lobe will be depressing one valve and spring to some extent. To prevent uneven stress to the camshafts when removing the camshaft bearing caps, Suzuki recommend that a large self-grip wrench be used to hold the camshaft down. A G-clamp of suitable size will make a substitute if a wrench is not available. Fit the wrench as shown

in the accompanying photograph, ensuring that it is so placed that slipping is not possible. Loosen evenly the four bolts holding each of the two cam bearing caps and then remove the bolts and caps. Note that each cap is marked, A, B, C or D. The casing below each cap is marked similarly, to enable the caps to be refitted in their original locations and positions. Also note the fitting of two small locating dowels to each of the cam bearing caps. Displace the clamping tool to free the camshaft.

6 If a suitable clamping tool is not available, it is acceptable to slacken the bearing cap bolts without restraining the camshafts, providing extreme caution is exercised. The bolts must be loosened evenly a little at a time, so that neither the camshaft nor the bearing caps are allowed to tilt.

7 Lift the cam chain off the sprocket and remove the camshaft, complete with the sprocket. Repeat the procedure for the other camshaft. Removal of the sprockets is not required unless the components require renewal.

8 If the top-end overhaul only is anticipated, the cam chain should not be allowed to fall down into the chain tunnel, as retrieval can be very difficult. Insert a long bar through the chain so that it rests on the cylinder head or use a length of stout wire secured to an adjacent stud or bolt hole.

7.1 Exercise care when separating cam cover from cylinder head

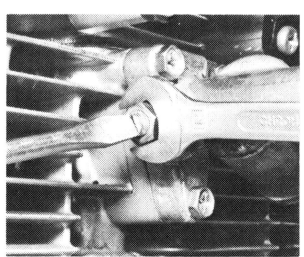

7.3 Slacken locknut and tighten plunger locking screw before removing cam chain tensioner unit

7.4 Jockey sprocket is held by four bolts

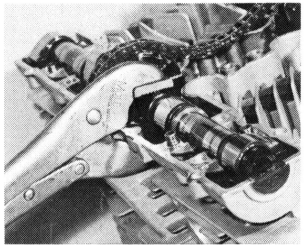
7.5 Suzuki recommend this method of camshaft retention

8 Dismantling the engine unit: removing the cylinder head, cylinder block and pistons

1 Do not disturb the cam followers and adjusters at this juncture. These components should be left in place until the examination stage. A useful method of ensuring these components stay in place, is to place a length of wood along the tops of these components, in the position of the camshafts; something similar to a section of broom handle would be ideal. Then replace one bearing cap on each wooden "camshaft" and lightly bolt it in position. The cylinder head is retained by twelve 10 mm nuts and three 6 mm bolts. Slacken and remove the 6 mm bolts from each end of the cylinder head (below the sparking plugs of cylinders No 1 and 4) and remove the third 6 mm bolt from the front of the cam chain tunnel. With the bolts removed, slacken the twelve nuts (8 have ordinary hexagon heads and 4 are chromed dome nuts) in the **reverse** order to that given in Fig. 1.21 which accompanies Section 52. To aid correct dismantling and rebuilding, the number of the nut is stamped next to it in the head casting. The nuts must be loosened evenly in this sequence, to avoid stressing the cylinder head casting.

2 Separate the cylinder head from the gasket, using a soft-headed mallet. Strike only those portions of the casing which are adequately supported, taking special care not to damage the

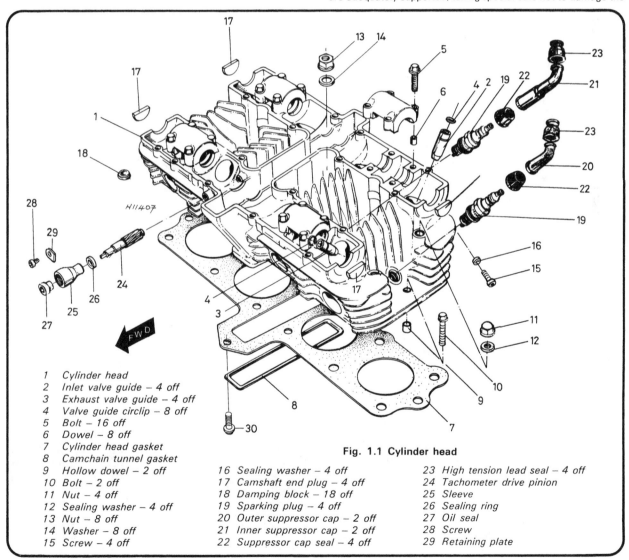

1 Cylinder head
2 Inlet valve guide – 4 off
3 Exhaust valve guide – 4 off
4 Valve guide circlip – 8 off
5 Bolt – 16 off
6 Dowel – 8 off
7 Cylinder head gasket
8 Camchain tunnel gasket
9 Hollow dowel – 2 off
10 Bolt – 2 off
11 Nut – 4 off
12 Sealing washer – 4 off
13 Nut – 8 off
14 Washer – 8 off
15 Screw – 4 off
16 Sealing washer – 4 off
17 Camshaft end plug – 4 off
18 Damping block – 18 off
19 Sparking plug – 4 off
20 Outer suppressor cap – 2 off
21 Inner suppressor cap – 2 off
22 Suppressor cap seal – 4 off
23 High tension lead seal – 4 off
24 Tachometer drive pinion
25 Sleeve
26 Sealing ring
27 Oil seal
28 Screw
29 Retaining plate

Fig. 1.1 Cylinder head

fins. Under no circumstances should levers be used between the mating surfaces of the cylinder head and cylinder block in an effort to facilitate separation. This action will lead to damage to the faces with subsequent risk of leakage. When lifting the cylinder head from position, the cam chain must be guided through the central tunnel and prevented from falling free. An extra pair of hands is beneficial at this stage. Before raising the cylinder head from position, remove the forward cam chain guide blade; this guide blade lifts out.

3   Separate the cylinder block from the base gasket, using the technique described for the cylinder head. Once again, levers should not be used. Lift the cylinder block upwards along the holding down studs until the piston skirts are visible but the piston rings still obscured by the cylinder bore spigots. If a top-end overhaul is to be carried out, the crankcase mouths must be padded with clean rags to prevent pieces cf broken piston from falling into the crankcase. The padding will also prevent the ingress of foreign matter during further dismantling or work. With the padding in place, lift the cylinder block off the pistons as squarely as possible to prevent the pistons from tying in the bore. Catch the pistons as they emerge from each bore, to prevent damage occurring. With the cylinder block removed, note the positioning of the two locating dowels and the use of O-rings on each of the outermost rear holding down studs.

4   Before removing the pistons, each should be marked using a metal scribe on the inside of the skirt. Number the pistons from 1 to 4, so that if they are to be re-used they may be fitted in their original locations. An arrow mark cast on each piston crown indicates the correct position in which the piston should be replaced.

5   Remove both circlips from each piston boss and discard them. Circlips should never be re-used if risk of displacement is to be obviated.

6   Using a drift of the correct diameter, tap each gudgeon pin out of the piston bosses until the piston can be lifted off the connecting rod, complete with rings. Make sure the piston is supported during this operation, or there is risk of bending the connecting rod.

7   If the gudgeon pin is a tight fit, do not resort to force. Warm the piston by placing a rag soaked in hot water on the crown, so that the piston bosses will expand and release their grip on the pin.

8   Each piston is fitted with three piston rings; two compression and one oil scraper. To remove the rings, spread the ends sufficiently with the thumbs to allow each ring to be eased from its groove and lifting clear of the piston. This is a very delicate operation necessitating great care. Piston rings are very brittle and will break easily. An alternative method of removing piston rings safely, especially when the rings are gummed up, is shown in the accompanying illustration. Use three narrow strips of tin cut from an old oil can, slipped between the rings and the piston.

8.1 A useful method of retaining cam followers and adjusters

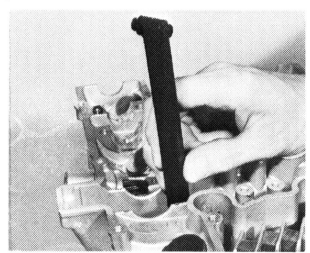
8.2 Lift out the forward cam chain guide blade

8.3 Lift the cylinder block off the pistons

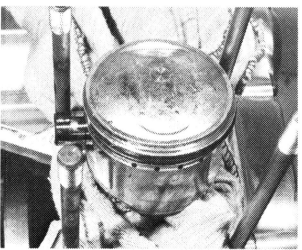
8.6 Displace the piston circlips and push out the gudgeon pin to free the piston

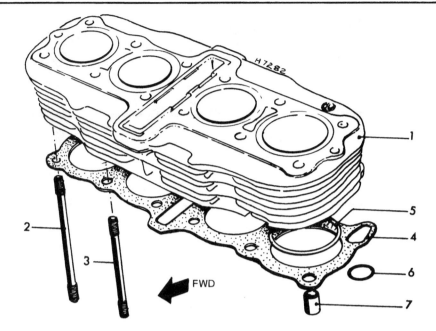

Fig. 1.2 Cylinder barrel

1   Barrel
2   Outer stud – 4 off
3   Inner stud – 8 off
4   Cylinder base gasket
5   O-ring – 4 off
6   O-ring – 2 off
7   Locating dowel – 2 off

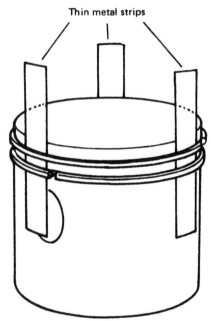

Thin metal strips

Fig. 1.3 Freeing gummed rings

9 Dismantling the engine unit: removing the contact breakers and automatic timing unit – GS 850 GN model

1   Free the contact breaker wiring leads from the clips positioned on the underside of the engine and displace the grommet from the casing wall. Place a spanner on the engine turning hexagon and loosen the contact breaker cam centre bolt using a second spanner. Remove the bolt and also the hexagon.
2   Before removing the contact breaker main baseplate, mark the relative position of the plate to the main casing, using either a centre punch or a screwdriver to make a mark on the plate that continues onto a web of the casing (see accompanying photograph). This is important, because on reassembly, the marks can be realigned, simplifying the ignition timing checking and adjustment operation. Unscrew the three screws which pass through the elongated holes in the main base plate

periphery. Detach the contact breaker assembly as a complete unit and remove the ignition timing index plate which is positioned behind the base plate. Lift the automatic advance unit (ATU) from position, noting the drive pin in the crankshaft end which locates with a recess in the rear of the ATU. Check the fit of the drive pin in the crankshaft, removing it if loose, to avoid accidental loss.
3   At this stage, the neutral indicator and gear position indicator switch and its lead can be detached. The switch is to be found inset below the secondary drive gear housing. Remove the two retaining screws and detach the switch unit. Care should be taken not to lose the switch contact and its small spring.

10 Dismantling the engine unit: removing the ignition signal generating unit and automatic timing unit – GS 850 GT/GLT models

1   Free the signal generating unit wiring leads from the clips positioned on the underside of the engine and displace the grommet from the casing wall. Place a spanner on the engine turning hexagon, and loosen the centre bolt. Remove the bolt and also the hexagon.
2   There is no necessity to mark the signal generating unit mounting plate before it is removed. If, however, reassembly is not to take place for some time, a scratch or punch mark on the mounting plate edge may provide a useful aid to the memory.
3   Unscrew the three screws which pass through holes in the mounting plate edge. Detach the signal generating assembly as a complete unit with its mounting plate. Also remove the ignition timing index plate which is fitted behind the mounting plate. Lift the automatic timing unit (ATU) from position, noting the drive pin in the crankshaft end which locates with a notch on the end face of the ATU. Check the fit of the drive pin in the crankshaft, removing it if loose, to avoid accidental loss.
4   Remove the neutral indicator/gear position indicator switch by following the procedure in paragraph 3 of the previous Section.

9.2a Punch mark contact breaker stator plate before ...

9.2b ... unscrewing three screws in elongated slots in baseplate edge and removing baseplate (GN model)

10.3 Remove drive pin if loose, to avoid loss

11 Dismantling the engine unit: removing the starter motor and intermediate gear

1   Remove the two screws which pass through the starter motor end cap flange into the crankcase. Ease the starter motor towards the right-hand side of the engine until the motor boss leaves the casing. Lift the starter motor up at the rear and out of the compartment. If the starter motor boss is tight in the casing, insert a wooden lever between the front of the motor and the casing wall. Use the lever to push the motor from place.
2   Free the alternator wiring lead from the guideway in the top of the crankcase. On removal of the alternator cover the lead must be fed through the casing wall because the leads are connected to the alternator stator. Loosen evenly and remove the alternator cover retaining screws. Lift the cover away and pull the lead through. Pass the connectors through the aperture individually.
3   Withdraw the intermediate gear spindle and lift the spindle and gear from position. Note the two shims, one of which is fitted to the spindle either side of the gear.

11.2 Lift off the alternator cover and feed the wires through casing wall

11.3 Withdraw spindle and remove intermediate gear

12 Dismantling the engine unit: removing the alternator, rotor and starter clutch

1   Loosen and remove the alternator rotor centre bolt after applying a spanner to the two flats on the rotor centre boss to prevent rotation. The 17 mm rotor centre bolt is extremely tight and its removal is further hampered by the use of substantial quantities of thread locking compound. The rotor is a very tight fit on the tapered crankshaft end and will require pulling from position. The rotor boss is threaded internally to take a slide hammer. This tool consists of a headed shaft upon which is placed a free sliding steel weight. After screwing the shaft into the centre boss the steel weight is slid forcibly along the shaft away from the rotor until it contacts the shaft head. The force imparted should release the taper. If the correct slide hammer is not readily available (Suzuki Service tool No. 09930-30102), an alternative method of drawing the rotor off its taper must be devised. A conventional two or three-legged puller can be used to good effect, noting that if the puller legs are large on the three-legged version it would probably not fit. When using this method, insert the centre bolt before fitting the puller. The bolt should be screwed in lightly, so that the centre of the puller is tightened down on to the bolt head, and cannot, therefore,

damage the internal thread on the crankshaft end. Under no circumstanced should levers be used in an attempt to remove the rotor. Such an approach will almost certainly lead to damage of the rotor or to the adjacent casings. When using the legged puller method of removal, the centre of the puller should be struck when it is tightened down; **Do not** strike the rotor itself as this could lead to damage and demagnetisation of the rotor magnets.

2   It may be found that the alternator rotor resists all attempts at removal. If this is the case, expert advice should be sought because the pressed up crankshaft and the rotor itself may suffer severe damage if excess force is employed. Provided that the rotor does not require attention and that the starter clutch and crankshaft oil seal are in good condition, these two components may be left on the crankshaft as their presence does not obstruct further dismantling. If work on the crankshaft and bearings is envisaged, the crankshaft will in any event require returning to a Suzuki Service Agent who may be entrusted to remove the alternator at the same time.

3   To continue dismantling, pull the rotor off the shaft complete with the starter clutch to which it is attached. Slide the large starter motor clutch gear off the crankshaft end and remove carefully the two needle roller bearing races and the large notched thrust washer.

12.1a Place spanner on rotor centre boss as bolt is loosened

12.1b A conventional legged puller may be used if slide hammer is unavailable

12.3 Slide off the rotor and starter clutch as a unit; note the two needle roller races

13 Dismantling the engine unit: removing the kickstart return spring – GS 850 GN model only

1   Remove the kickstart lever from the splined shaft after unscrewing fully the pinch bolt. Loosen evenly and remove the ten crosshead screws which secure the clutch/primary drive cover on the right-hand side of the engine. If necessary, use a soft-headed mallet to release the cover from the gasket.

2   Using a large pair of pliers, grasp the outer turned end of the kickstart return spring and pull it from the anchoring recess in the casing. Allow the spring to unwind in a controlled manner. Withdraw the spring guide and then free the spring by displacing the inner turned end from the radially drilled hole in the shaft. Pull the spring from position.

14 Dismantling the engine unit: removing the clutch

1   Loosen evenly and remove the ten crosshead screws which secure the clutch/primary drive cover on the right-hand side of the engine. Remove the cover and gasket, tapping the cover with a soft-headed mallet if necessary to break the joint with the gasket.

2   Unscrew the six clutch pressure spring bolts evenly and remove them, together with the springs. Remove the clutch pressure plate and then withdraw the clutch plates one at a time, noting the alternating sequence. Pull the clutch operating thrust piece (release rack), the needle roller thrust bearing and the washer from the pressure plate. Ensure these small components are not mislaid. Note that the innermost plain plate is secured by a thin spring band which need not be disturbed at this juncture.

3   Bend down the 'ear' of the tab washer which secures the clutch centre nut. The nut may be very tight, and in order to prevent the shaft rotating, adopt the following procedure. Fabricate a locking sprag from a piece of plate steel and position it between one of the teeth on the clutch centre boss, and the inside of the clutch outer casing. To prevent damage to the outer casing a thin length of wood should be positioned between the steel plate and the casing. Care must be taken when carrying out this operation that the steel plate does not slip off the clutch centre boss tooth that it is secured against. An alternative to this method is to temporarily refit the gear change lever and place the transmission in top gear. The output shaft flange should now be prevented from turning. This may be accomplished with ease by inserting two of the four bolts in

diametrically opposed holes in the flange and then placing a long bar across the bolt heads. The nut can now be removed followed by the tab washer. Whichever method is used, common sense should prevail in the amount of force used, and care should be taken not to damage any of the components.

4   Withdraw the clutch centre boss and remove the thrust washer from the shaft. The clutch outer drum revolves on a caged needle roller bearing, supported on a large central spacer. To enable the edge of the outer drum to clear the casing on removal, the spacer must be withdrawn. Use a suitable crankcase cover retaining screw, screwed into one of the threaded holes in the spacer as a means of removal. Pull out the needle bearing and then lift the outer drum towards the rear and out of the primary drive case. The oil pump drive gear fitted to the rear of the clutch may now be removed together with the needle roller bearing, spacer and backing washer. When removing the clutch centre boss, note the fitting of a large wave washer and seat around the base flange of the boss. The wave washer and its seat will be partially obscured by the last remaining plain plate. The same spring steel band that secures the innermost plate also secures the wave washer and seat. The reason for the fitting of the wave washer is to aid in smooth clutch engagement.

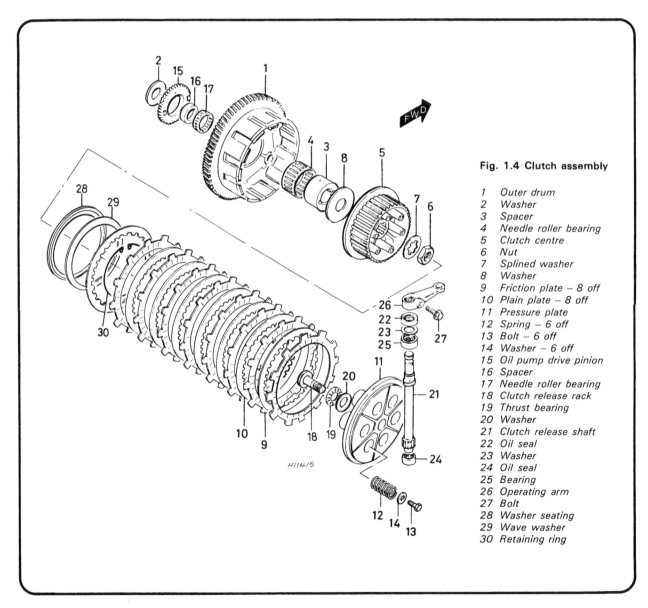

Fig. 1.4 Clutch assembly

1   Outer drum
2   Washer
3   Spacer
4   Needle roller bearing
5   Clutch centre
6   Nut
7   Splined washer
8   Washer
9   Friction plate – 8 off
10 Plain plate – 8 off
11 Pressure plate
12 Spring – 6 off
13 Bolt – 6 off
14 Washer – 6 off
15 Oil pump drive pinion
16 Spacer
17 Needle roller bearing
18 Clutch release rack
19 Thrust bearing
20 Washer
21 Clutch release shaft
22 Oil seal
23 Washer
24 Oil seal
25 Bearing
26 Operating arm
27 Bolt
28 Washer seating
29 Wave washer
30 Retaining ring

13.2 Remove the kickstart spring guide and spring (GN model)

14.2a Remove the clutch pressure plate springs and bolts and...

14.2b ... the pressure plate and clutch plates

14.4a After centre nut and tab washer removal, pull centre boss off shaft

14.4b Displace central spacer and bearing to allow ...

14.4c ... removal of the clutch drum/primary driven gear

14.4d Remove the oil pump drive gear, spacer and bearing

15 Dismantling the engine unit: removing the gear selector external components and the oil pump

1   Grasp the gearchange shaft at the quadrant end and withdraw it from the casing, complete with the centraliser spring. Using an impact driver, remove the two countersunk crosshead screws from both the change drum guide plate and the panel operating plate. These are the two lowest plates in the casing, positioned to the left and right respectively of the selector quadrant. Pinch together the two spring loaded pawls which are fitted to the selector quadrant, and withdraw the quadrant, complete with pawls, from the end of the change drum. Store the pawls and springs safely to avoid loss. Removal of all these components can be carried out at this stage or later, after separation of the crankcase halves.
2   To remove the oil pump, the oil pump driven gear must first be removed. Displace the circlip which retains the gear and slide the gear off the pump drive pin. Note the fitting of a washer, which should also be removed, behind the driven gear.
3   Remove the three screws which retain the oil pump in position and pull the oil pump from place. These crosshead screws will almost certainly again be very tight, and require removal with an impact driver. With these screws removed, pull the oil pump from place. Note the two O-rings fitted to the rear of the pump; these should be renewed as a matter of course upon reassembly.
4   There are now three remaining plates in the casing to be detached. Remove the mainshaft bearing retainer plate, which is retained by three countersunk screws to the right of the mainshaft. Then remove the oil passage plate, similarly retained by three countersunk screws, from above the mainshaft. This plate may well need lightly prising from position due to its sealing gasket forming a strong joint. The third plate, held by four countersunk screws, to the left of the mainshaft, is the layshaft lubrication end plate. When removing this plate, the sealing gasket should come free at the same time. Note that removal of the oil passage plate, whilst not being strictly necessary, can be useful to check for clogging etc. As with all the previous screws of this type, the countersunk screws which secure these plates will probably be very tight. An impact driver should be used to facilitate removal.

16 Dismantling the engine unit: removing the sump and oil strainer

1   Invert the engine so that the sump is facing upwards and the crankcase upper half is resting on its rear edge and on the cylinder holding down studs, noting the positioning of the cable guide clip. Lift the sump away after releasing it from the gasket, using a soft-headed mallet if necessary.

2   The oil strainer is retained on the oil pick-up chamber by three crosshead screws. Remove the screws and lift the screen away.

17 Dismantling the engine unit: removing the secondary drive units and separating the crankcase halves

1   Place the engine the normal way up on the bench and evenly slacken and remove the sixteen crankcase top half securing bolts; after engine no 130760 (later GT and GLT models), note that four of the 8 mm bolts are secured by nuts, two of which have separate washers. Invert the engine again so that it is resting on the cylinder holding studs and the rear edge of the upper casing. Loosen evenly and remove the eleven 6 mm and thirteen 8 mm securing bolts. Each 8 mm bolt is identified by a number stamped on an adjoining portion of the casing. To prevent stress in the casings, these bolts should be slackened in sequence, commencing with the highest number and continuing through to the lowest number. Also note that the No 6 bolt has the alternator lead guide clip attached to it.
2   The engine can now be left in this position, so that on separation of the crankcase halves, the lower half is lifted away leaving the crankshaft, gearbox components and kick starter shaft (where fitted) in the upper crankcase half. Because the gear change internal mechanism – the gear selector drum and forks – are fitted to the lower casing, these components will be lifted away in situ as the lower half casing is separated.
3   Separation of the crankcase halves should be carried out with care, using a soft-headed mallet initially, to release the two cases from the gasket compound which was used on original assembly. **Do not** use levers placed between the two mating surfaces in an effort to hasten separation. Treatment of this nature will almost certainly damage the machined surfaces, causing subsequent oil leakage.
4   Before attempting actual separation of the two crankcase halves, remove the secondary drive and driven bevel gear units. Remove the four mounting bolts that pass through the secondary drive gear housing flange into the left-hand side of the engine. Similarly, remove the four mounting bolts from the circular secondary driven gear housing flange at the rear of the engine. The two units can now be removed from the crankcase – they will pull free. If either unit resists removal, continue with the crankcase separation operation and free the secondary drive/driven units when, with a reduction in the clamping pressure, they can be prised from the cases. Note the large shims on each housing; ensure these are not discarded or misplaced.
5   After separation, study the internal components carefully before continuing with the dismantling operation. This will help prevent confusion when reassembly is being carried out. Note and remove the O-ring which seats in a recess in the upper

15.3 Oil pump is held to rear of casing by three screws

16.2 Remove the three screws and lift the oil strainer screen away

17.4a Remove the secondary drive bevel gear unit ...

17.4b ... followed by the driven gear unit from the rear of the engine

17.5 Main components remaining in the upper casing after separation

casing. Check the two hollow location dowels for tightness. If loose, they should be removed, to avoid loss.

18 Dismantling the engine unit: removing the crankshaft, gear shafts and kickstart shaft

1   Grasp the crankshaft with both hands and lift it upwards out of the casing, as a complete unit, together with the oil seals and the cam chain. If the crankshaft is firmly seated, use a soft-headed mallet to free it from the casing. Note the five crankshaft main bearing outer race locating pins in the upper crankcase half. If they are loose, remove them using a pair of long nose pliers.

2   Lift out the two gearbox shafts individually, complete with pinions and seals, and the mainshaft spring damper unit. Note the positions of the bearing location half, or C-clips, which prevent axial movement, and prise them from position. The two shaft assemblies should be put to one side for further attention at a later stage.

3   Lift out the kickstart shaft (GN model only) and place it to one side for examination later. Note the two small locating pins for the shaft; if they are loose remove them with long-nose pliers to avoid loss.

18.3 Lift out the kickstart shaft (GN model)

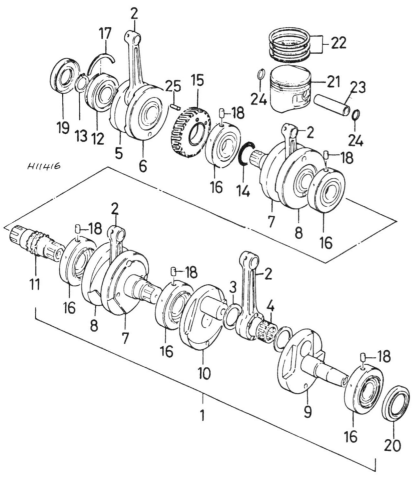

H11416

Fig. 1.5 Pistons and crankshaft assembly

1   *Crankshaft*
2   *Connecting rod – 4 off*
3   *Thrust washer – 4 off*
4   *Needle roller bearing – 4 off*
5   *Flywheel*
6   *Flywheel*
7   *Flywheel/mainshaft*
8   *Flywheel/crankpin*
9   *Flywheel/mainshaft*
10 *Flywheel/crankpin*
11 *Mainshaft/cam sprocket*
12 *Right-hand bearing*
13 *Circlip*
14 *Thrust washer*
15 *Primary drive gear*
16 *Bearing – 5 off*
17 *Bearing locating ring*
18 *Pin – 5 off*
19 *Right-hand oil seal*
20 *Left-hand oil seal*
21 *Piston – 4 off*
22 *Piston ring set – 4 off*
23 *Gudgeon pin – 4 off*
24 *Circlip – 8 off*
25 *Pin – 5 off*

20 Examination and renovation: general

1   Before examining the component parts of the dismantled engine/gear unit for wear, it is essential that they should be cleaned thoroughly. Use a paraffin/petrol mix to remove all traces of oil and sludge which may have accumulated within the engine.

2   Examine the crankcase castings for cracks, or other signs of damage. If a crack is discovered, it will require professional attention, or in an extreme case, renewal of the casting.

19 Dismantling the engine unit: removing the gear selector internal mechanism

1   Withdraw the selector fork rods towards the primary drive side of the engine and displace the selector forks. There are two rods, of which the front carries one fork and the rear two forks.

2   The cam stopper arm is contained within the gearbox casing, pivoting on the forward rod. The arm return spring is hooked over the rear rod.

3   Remove the neutral position drum detent housing bolt and displace the detent spring and plunger. The change drum can now be pulled out of the casing towards the primary drive side of the engine.

3   Examine carefully each part to determine the extent of wear. If in doubt, check with the tolerance figures whenever they are quoted in the text. The following Sections will indicate

what type of wear can be expected and in many cases, the acceptable limits.

4   Use clean, lint-free rags for cleaning and drying the various components, otherwise there is risk of small particles obstructing the internal oilways.

19.1a Remove the plunger pawl lifter plate and ...

19.1b ... the drum guide plate

19.3 With the selector fork rods removed, withdraw the change drum from the casing

20.3 Displace main bearing outer races to make visual inspection of bearings

20.5 Renew the crankshaft end oil seals upon reassembly

21 Crankshaft assembly: examination and renewal

1   The crankshaft assembly comprises four separate sets of flywheels, with their respective big-ends, connecting rods and main bearings, pressed together to form a single unit.
2   Due to the complex construction of the crankshaft, damaged crankshafts are not easily repaired. In the event of main bearing or big-end bearing failure a new crankshaft must be acquired; Suzuki provide a service exchange scheme through which rebuilt crankshafts can be supplied.
3   Main bearing failure will immediately be obvious when the bearings are inspected after the old oil has been washed out. If any play is evident or if the bearings do not run freely, renewal is essential. The main bearings are of the caged roller type and as the outer races are not restrained axially they may be displaced to one side, to aid visual inspection. Warning of main bearing failure is usually given by a characteristic rumble that can be readily heard when the engine is running. Some vibration will also be felt, which is transmitted via the footrests,
4   Big-end failure is characterised by a pronounced knock that will be most noticeable when the engine is working hard. There should be no play whatsoever in any of the connecting rods,
when they are pushed and pulled in a vertical direction. Check also the deflection of each connecting rod in line with the crankshaft, taking the measurement as the small-end eye. Movement greater than 3 mm (0.118 inch) indicates a worn big-end bearing. Using a feeler gauge, check the axial side play at the big-ends. The total movement should be no greater than 1.00 mm (0.040 inch).
5   The oil seals at each end of the crankshaft are easy to renew when the engine is stripped; they are a push fit over each end of the crankshaft, one against and the other close to the outer main bearings. It is a wise precaution to renew these seals whenever the engine is stripped, irrespective of their condition.

22 Connecting rods: examination and renovation

1   It is unlikely that any of the connecting rods will bend during normal usage, unless an unusual occurence such as a dropped valve has caused the engine to lock. Carelessness when removing a tight gudgeon pin can also give rise to a similar problem. It is not advisable to straighten a bent connecting rod; renewal is the only satisfactory solution.

2   The small-end eye of each connecting rod is unbushed and it will be necessary to renew the connecting rod if the gudgeon pin becomes a slack fit due to wear in the eye rather than the pin. Refer to the advice about crankshaft renewal given in the previous Section. Check the clearance using the unworn end of a gudgeon pin, as the centre portion of a used pin will by slightly worn. The maximum clearance should not exceed 0.12 mm (0.0047 in). Always check that the oil hole in the small-end eye is not blocked since if the oil supply is cut off, the bearing surfaces will wear very rapidly.

23 Cylinder block: examination and renovation

1   The usual indication of badly worn cylinder bores and piston is excessive smoking from the exhausts and piston slap; a metallic rattle that occurs when there is little or no load on the engine. If the top of the bore of the cylinder block is examined carefully, it will be found that there is a ridge on the thrust side, the depth of which will vary according to the amount of wear that has taken place. This marks the limit of travel of the uppermost piston ring.
2   Measure the bore diameter just below the ridge. Take two measurements, at 90° to one another. Take two similar measurements half way down the bore and at a position just above the lower edge of the bore. If any measurement exceeds the maximum allowable, the cylinder should be rebored and fitted with an oversize piston. If the difference between the maximum and minimum measurement exceeds 0.085 mm (0.0035 in) a rebore is also required.
3   If an internal micrometer is not available, the amount of cylinder bore wear can be approximated by inserting the piston without rings so that it is approximately $\frac{3}{4}$ inch from the top of the bore. It it is possible to insert a 0.060 mm (0.0024 in) feeler gauge between the piston and the cylinder wall on the thrust side of the piston, remedial action must be taken.
4   Oversize pistons are available in two sizes: + 0.5 mm (0.020 inch) and + 1.0 mm (0.040 in)
5   Check that the surface of the cylinder bores is free from score marks or other damage that may have resulted from an earlier engine seizure or a displaced gudgeon pin. A rebore will be necessary to remove any deep scores, irrespective of the amount of bore wear that has taken place, otherwise a compression leak will occur.
6   Make sure the external cooling fins of the cylinder block are not clogged with oil or road dirt, which will prevent the free flow of air and cause the engine to overheat.

24 Pistons and piston rings: examination and renovation

1   Attention to the piston and piston rings can be overlooked if a rebore is necessary, since new components will be fitted.
2   If a rebore is not considered necessary, examine each piston closely. Reject pistons that are scored or badly discoloured as the result of exhaust gases by-passing the rings.
3   Remove all carbon from the piston crowns, using a blunt scraper, which will not damage the surface of the piston. Clean away all carbon deposits from the valve cutaways and finish off with metal polish so that a clean, shining surface is achieved. Carbon will not adhere so readily to a polished surface. Using an external micrometer or vernier gauge, measure the external diameter of each piston across the thrust faces (at 90° to the gudgeon pin line), at the bottom of the skirt. Take a second measurement approximately 15 mm (0.590 in) up from the lower edge of the skirt. If either measurement is less than that given for the service limit 68.880 mm (2.7118 in), the piston is in need of renewal.
4   Check that the gudgeon pin bosses are not worn or the circlip grooves damaged. Check that the piston ring grooves are not enlarged. Side float should not exceed 0.18 mm (0.007 in) for the top ring and 0.15 mm (0.006 in) for the second ring.

5   Piston ring wear can be measured by inserting the rings in the bore from the top, pushing them down with the base of the piston so that they are square in the bore and about $1\frac{1}{2}$ inches down. If the end gap exceeds 0.7 mm (0.03 in) in any of the rings, renewal is necessary. A replacement set of rings is comparatively inexpensive and it is considered good practice to renew that as a matter of course whenever the engine is dismantled.
6   Check that there is no build-up of carbon on the inside surface of the rings or in the grooves of the pistons. Any build-up should be removed by careful scraping.
7   The piston crowns will show whether the engine has been rebored on some previous occasion. All oversize pistons have the rebore size stamped on the crown. This information is essential when ordering replacement piston rings.
8   If new piston rings are fitted, but a rebore has not taken place, the cylinder bores should be 'glaze busted'. This honing operation, as the name suggests, remove the glazed surface of the bore which has been caused by the countless up and down strokes of the piston and rings. If glaze busting is not carried out, the time required to run-in the new rings will be greatly, and unnecessarily extended.

24.2 Examine pistons for this type of scoring

25 Examination and renovation: cylinder head and valves

1   Remove the cam followers and adjuster shims from the cylinder head, marking each follower so that it may be refitted in its original location. It is best to remove all carbon deposits from the combustion chambers before removing the valves for inspection and grinding-in. Use a blunt end chisel or scraper so that the surfaces are not damaged. Finish off with a metal polish to achieve a smooth, shining surface. If a mirror finish is required, a high speed felt mop and polishing soap may be used. A chuck attached to a flexible drive will facilitate the polishing operation.
2   A valve spring compression tool must be used to compress each set of valve springs in turn, thereby allowing the split collets to be removed from the valve collar, and the valve springs and collars to be freed. Keep each set of parts separate and mark each valve so that it can be replaced in the correct combustion chamber. There is no danger of inadvertently replacing an inlet valve in an exhaust position, or vice-versa, as the valve heads are of different sizes. The normal method of marking valves for later identification is by centre punching them on the valve head. This method is not recommended on valves, or any other highly stressed components, as it will

produce high stress points and may lead to early failure. Tie-on labels, suitable inscribed, are ideal for the purpose.

3   Check the valve stems and guides for wear, either by direct measurement, or by inserting a valve and rocking it to and fro both along the direction of cam lobe thrust and at right angles to it; any wear will be most noticeable when the valve is at the maximum lift position. Compare the original valve with a new component to check if wear is excessive. If a small bore gauge and micrometer are available, the two components can be measured at three or four points along their bearing surfaces, both in the direction of cam lobe thrust and at right angles to it. Subtract the smallest stem diameter measurement obtained from the largest guide bore diameter; if the stem/guide clearance figure thus calculated exceeds the specified service limit, one or both components must be renewed.

4   To remove the old valve guide, place the cylinder head in an oven and heat it to about 150°C. The old guide can now be tapped out from the cylinder side. To prevent distortion of the large alloy casting it is essential that the cylinder head is heated evenly. For this reason an oven **must** be used in preference to a blow torch or other methods of heating. If inexperienced in this type of work, the advice of a Suzuki Service Agent should be sought. Before drifting a guide from the place, remove any carbon deposits which may have built up on the guide end projecting into the port. Carbon deposits will impede the progress of the guide and may damage the cylinder head. If possible, use a double diameter drift. The smaller diameter should be close to that of the valve stem, and the larger diameter slightly smaller than that of valve guide. Provided that care is exercised, a parallel shanked drift may be used as a substitute. New guides may be fitted by using the same procedure, after which the valve seats must be cut as described in paragraph 6.

5   Valve grinding is a simple task. Commence by smearing a trace of fine valve grinding compound (carborundum paste) on the valve seat and apply a suction tool to the head of the valve. Oil the valve stem and insert the valve in the guide so that the two surfaces to be ground in make contact with one another. With a semi-rotary motion, grind in the valve head to the seat, using a backward and forward action. Lift the valve occasionally, rotating it through about 45° so that the grinding compound is distributed evenly. Repeat the application until an unbroken ring of light grey matt finish is obtained on both valve and seat. This denotes the grinding operation is now complete. Before passing to the next valve, make sure that all traces of the valve grinding compound have been removed from both the valve and its seat and that none has entered the valve guide. If this precaution is not observed, rapid wear will take place due to the highly abrasive nature of the carborundum base.

6   If, after grinding, it is found that the width of the grey seating ring is greater than 1.5 mm (0.06 in) the valve seat must be recut using a special cutting tool. It will be seen from the accompanying illustration that angles of 75° and 15° must be cut in order to reduce the valve seat width, followed by a 45° cut in order to restore the correct seat angle and the correct seat width to within the range 1.0 – 1.2 mm (0.04 – 0.05 in). Because of the expense of purchasing the three seat cutters and because of the accuracy with which cutting must be carried out, it is strongly recommended that the cylinder head be returned to a Suzuki Service Agent for attention. It follows that when material is removed from the valve seat, the valve stem will protrude further from the upper side of the cylinder head. In extreme cases it may be found that on adjustment of the cam clearances, the prescribed clearance cannot be arrived at even with the thinnest adjustment shim available. If this is found to be the case, removal of a small amount of metal from the valve stem end is permissible. Grinding should be carried out on a suitable machine so that the stem end remains square with the shank. Where grinding to attain the correct clearance reduces the distance between the top of the stem and the upper edge of the collet groove to less than 4.00 mm (0.1574 in) a new valve seat insert must be fitted. This operation is highly skilled requiring the use of very specialised equipment.

7   Where deep pitting of the seat and valve is encountered the seat should be recut as previously described. The valve face may be ground back on a special grinding machine to an angle of 45°, provided that after grinding, the depth of the valve periphery has not been reduced to less than 0.5 mm (0.0197 in).

8   Examine the condition of the valve collets and the groove on the valve stem in which they seat. If there is any sign of damage, new parts should be fitted. Check that the valve spring collar is not cracked. If the collets work loose or the collar splits whilst the engine is running, a valve could drop into the cylinder and cause extensive damage.

9   Check the free length of each of the valve springs. The springs have reached their serviceable limit when they have compressed to the limit readings given in the Specifications Section of this Chapter.

10 Reassemble the valve and valve springs by reversing the dismantling procedure. Ensure that all the springs are fitted with the close coils downwards towards the cylinder head. Fit new oil seals to each valve guide and oil both the valve stem and the valve guide, prior to reassembly. Take special care to ensure the valve guide oil seal is not damaged when the valve is inserted. As a final check after assembly, give the end of each valve stem a light tap with a hammer, to make sure the split collets have located correctly.

11 Check the cylinder head for straightness, especially if it has shown a tendency to leak oil at the cylinder head joint. If there is any evidence of warpage, provided it is not too great, the cylinder head must be either machined flat or a new head fitted. Most cases of cylinder head warpage can be traced to unequal tensioning of the cylinder head nuts and bolts by tightening them in incorrect sequence.

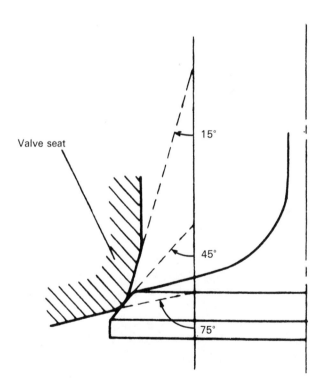

Fig. 1.6 Valve seat grinding angles

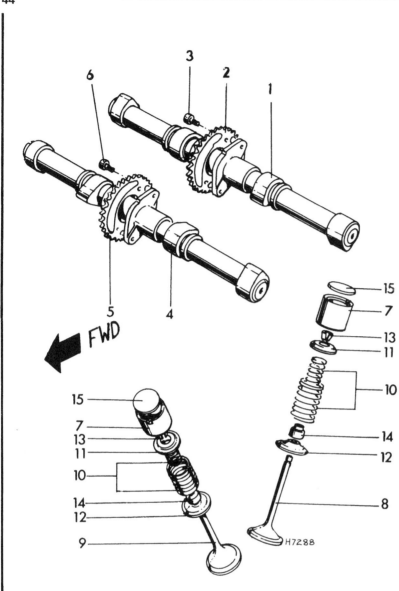

FWD

Fig. 1.7 Camshafts and valves

1   Inlet camshaft
2   Inlet camshaft sprocket
3   Bolt – 2 off
4   Exhaust camshaft
5   Exhaust camshaft sprocket
6   Bolt – 2 off
7   Cam follower – 8 off
8   Inlet valve – 4 off
9   Exhaust valve – 4 off
10 Spring set – 8 off
11 Upper spring seat – 8 off
12 Lower spring seat – 8 off
13 Valve collet – 16 off
14 Oil seal – 8 off
15 Shim

H7288

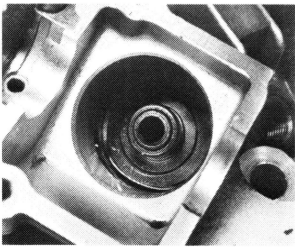

25.10a Fit new oil seals to each valve guide

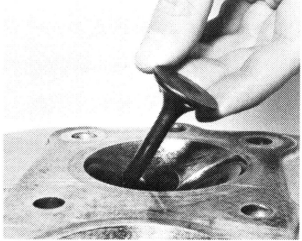

25.10b Lubricate the valve stem thoroughly before insertion

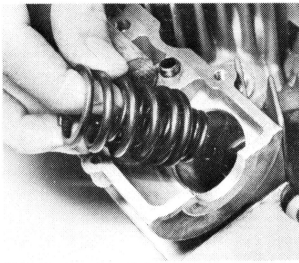

25.10c Fit valve springs with the close coils downwards and ...

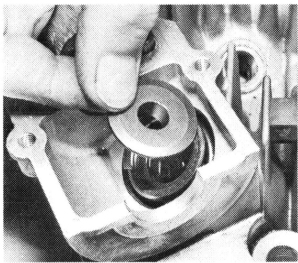

25.10d ... then place top spring seat in position, before ...

25.10e ... compressing the springs and refitting the split collets

25.10f Cam follower and adjustment shim ...

25.10g ... must be inserted **squarely**

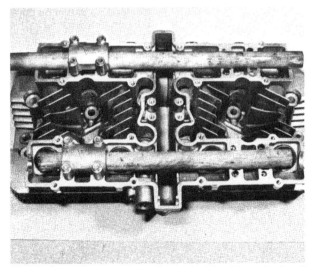

25.10h Cam followers etc may be retained by this method until reassembly

26 Examination and renovation: camshafts, cam followers and camshaft drive sprockets

1   Inspect the cams for signs of wear such as scored lobes, scuffing, or indentation. The cams should have a smooth surface. The complete camshaft must be replaced if any lobes are worn or indented, through lubrication failure etc. In due course even normal wear of each cam lobe may progress to the stage where full valve lift is no longer possible. Measure each cam from the lobe to the base circle, comparing the overall height with these figures.

Minimum cam lobe height:
  Inlet   *36.020 mm (1.4181 in)*
  Exhaust   *35.470 mm (1.3965 in)*

If any one cam on either camshaft is below the minimum figure, the camshaft should be renewed in order to restore performance.
2   Refit both camshafts in the cylinder head and fit the bearing caps and bolts. Tighten the bolts to a torque wrench setting of 1.0 kgf m (7.5 lbf ft). Check the clearance between the camshaft journals and the bearing surfaces. This is most easily accomplished by fitting a dial gauge to the cylinder head and moving the camshaft radially in a vertical or horizontal plane. If the clearances exceed those figures given in the Specifications section of this Chapter, remove the camshafts, refit the bearing caps and check the diameter of each bearing surface, to determine whether the camshaft or cam bearing surface is at fault.
3   If it is found that the camshaft bearings are worn or badly scored, the cylinder head and bearing caps must be renewed. There is no provision for renewing the bearing surfaces; the camshafts run directly in the cylinder head material and that of the bearing caps, there being no separate bearings.
4   Examine the camshaft chain sprockets for hooked, worn, or broken teeth. If any damage is found, the camshaft sprocket in question should be renewed. Each sprocket is retained on the camshaft flange by two socket screws. When refitting either sprocket note that each is marked IN or EX as are the camshafts. It is important that the sprockets are fitted on the correct camshaft and in the position shown in the accompanying illustration. Incorrect assembly will prevent accurate valve timing. Apply a small quantity of locking fluid to the securing screws during assembly.
5   The camshaft drive sprocket is an integral part of the crankshaft and therefore, if damage is evident, the complete crankshaft must be renewed. Fortunately, this drastic course of action is rarely necessary since the parts concerned are fully enclosed and well lubricated, working under ideal conditions.
6   Inspect the external surfaces of the cam followers for signs of scoring or fracture. If damage is evident, the component must be renewed. If scoring has occurred, it follows that similar

damage may be found in the appropriate guide tunnel in the cylinder head. Damage to the tunnels cannot be rectified under normal circumstances and therefore a new cylinder head must be obtained. Insert each cam follower into the guide tunnel from which it was removed. Only the lightest pressure should be used to insert each follower. If a follower is inserted even at a slight angle, binding against the tunnel will result. Any effort made to tap the follower in will almost certainly jam the follower solidly. Removal is then very difficult!
  Check the clearance between each cam follower and guide tunnel. Unfortunately no precise figures are available but the follower should be a good sliding fit, with no perceptible play from side to side. Excess play will allow the cam follower to tilt, causing noisy operation and accelerated wear of the cylinder head.

27 Examination and renovation: cam chain and chain tensioner mechanism

1   Inspect the cam chain for obvious signs of damage, such as broken or missing rollers or fractured links. Some indication of the amount of chain wear may be gained by checking the extent of adjustment remaining on the automatic tensioner assembly. If the plunger has moved towards the end of the stroke, it may be assumed that the chain is near the end of its useful life. Wear of the chain can be measured by washing it in petrol, then pulling on the chain, with one end anchored, so that it stretches as far as possible. If the length of 20 links exceeds 157.80 mm (6.213 in), the chain is past its working life, and must be renewed. Although the cam chain works in almost ideal conditions, being fully lubricated and enclosed, wear will develop after an extended mileage. If there is any doubt as to the chain's condition, it should be renewed, as breakage will cause extensive engine damage.
2   Loosen the locking screw on the chain tensioner body to free the plunger pushrod. Rotate the adjuster knob anti-clockwise so that the plunger may be pushed in fully, and check that the plunger moves in and out freely, without any tendency to bind. If plunger movement is not perfectly smooth, the complete unit should be renewed.
3   The cam chain is tensioned via a steel-backed rubber blade, by means of the aforementioned plunger device. In addition, there is a second blade at the front of the cam chain tunnel which acts as a guide. Inspect the surfaces of both blades. If the rubber has been badly scored by the chain or is coming away from the steel backing, the blade in question should be renewed.
4   The jockey sprocket fitted between the two camshafts should also be examined at this juncture. Clean the sprocket and carrier assembly thoroughly in petrol and check that the sprocket rotates freely. Inspect the sprocket teeth using criteria for renewal as given for the camshaft sprockets.

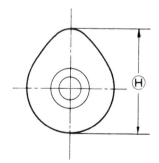

Fig. 1.8 Measurement of cam lobe height

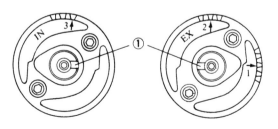

Fig. 1.9 Camshaft sprocket positioning marks

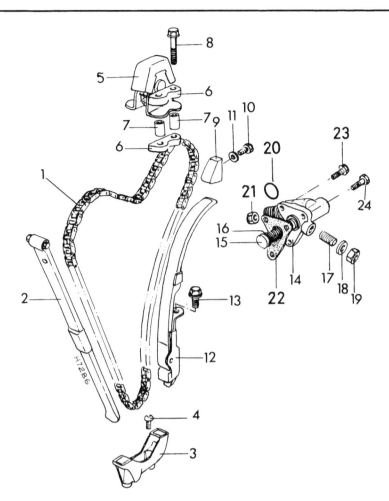

Fig. 1.10 Cam chain and tensioner

1   Cam chain
2   Chain guide blade
3   Blade retainer
4   Screw – 2 off
5   Jockey sprocket
6   Rubber damper – 4 off
7   Spacer – 4 off
8   Bolt – 4 off
9   Chain guide block
10 Bolt – 2 off
11 Washer – 2 off
12 Chain tensioner blade
13 Bolt
14 Tensioner adjuster assembly
15 Plunger
16 Spring
17 Screw
18 Washer
19 Nut
20 O-ring
21 Nut
22 Tensioner adjuster gasket
23 Bolt – 2 off
24 Bolt

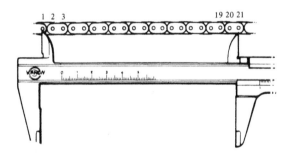

Fig. 1.11 Measuring cam chain wear

28 Examination and renovation: tachometer drive assembly

1   The worm drive to the tachometer is an integral part of the exhaust camshaft which meshes with a pinion attached to the cylinder head cover. If the worm is damaged or badly worn it will, unfortunately, be necessary to renew the camshaft complete.

2   The driveshaft and pinion are a single part retained in the cylinder head in a bush housing which is secured by a claw and screw. Renewal is therefore straightforward. It is unlikely that wear will develop on either the drive or driven pinion as both are well lubricated and lightly loaded.

29 Examination and renovation: gearbox components

1   It should not be necessary to dismantle either of the gear clusters unless damage has occurred to any of the pinions or if the journal ball bearings require attention.

2   The accompanying illustration shows how both clusters of the gearbox are assembled on their respective shafts. It is imperative that, the gear clusters, including the various types of washers, spacers and circlips, are assembled in EXACTLY the correct sequence, otherwise constant gear selection problems will occur.

3   In order to eliminate the risk of misplacement, make rough sketches as the clusters are dismantled. Also strip and rebuild as soon as possible to reduce any confusion which might occur at a later date.

4   As noted earlier in the general description section of this Chapter, there are three shafts within the gearbox. The mainshaft is of unusual construction in that it is in fact one shaft within another. These will be referred to as the inner and outer mainshaft in the text. The layshaft (driveshaft) is of normal construction.

4   If, on inspection, the bearings or gear pinions are damaged, the bearings must be pulled from place to allow renewal or further dismantling. A standard two or three-legged puller may be used to good effect on these bearings.

5   When dismantling the layshaft, certain points should be noted. The layshaft right-hand bearing is retained by a circlip. To remove this bearing, first displace the circlip and then draw off the bearing. There is a special dogged locking washer fitted to the layshaft between two adjacent gear pinions. The washer is

placed, along with another splined washer, between the 3rd gear and 5th gear driven pinions. (The 5th gear pinion can be identified easily because it is black; caused by the hardening treatment. To free the dogged splined washer, and so release the next washer in the sequence, the special washer must be turned slightly so that the internal serrations clear the shaft splines. The washer may then be pulled from place. Similarly, on reassembly, the special washer must be turned so that it engages and 'locks' correctly with the next washer. Also upon reassembly, ensure that the spacer which fits inside the centre of the 3rd gear driven pinion, is fitted so that the oil hole aligns with the oil hole in the layshaft. Note that the circlip that secures the 3rd gear pinion must be fitted with the chamfered side of the circlip towards the 3rd gear pinion as shown in the accompanying illustration. The layshaft left-hand bearing is located by a collar which is a tight interference fit in the shaft. Ideally the bearing and collar should be removed using a fly-press, but it may be possible to accomplish removal using a large two-or three-legged puller.

6   To dismantle the two-piece mainshaft assembly is slightly more difficult than the same operation on the layshaft. In order to separate the two sections of the mainshaft, the large spring shock absorber must first be displaced. This is not an easy task; the spring is heavy gauge and extremely powerful. The correct method of spring removal, is to draw off the left-hand outer mainshaft bearing followed by the small spacer, and then fit the Suzuki special service tool (Part No. 09924-44510) to effect spring removal The spring is retained in position in a similar way to a valve spring, albeit of somewhat inflated proportions, by the use of two collets on top of a large spring top seat. The Suzuki service tool compresses the spring sufficiently to allow the collets to be removed safely and the spring then slid off the shaft. This method of spring removal is obviously the safest, easiest and most effective way, and some effort to obtain the service tool, from a friendly local dealer on a temporary basis perhaps, may prove beneficial. If the Suzuki tool is not available, an alternative method of spring removal must be adopted. Approach with caution. Fabricate a U-shaped end piece of a correct size to fit the top of the spring seat, which will clear the protruding end of the shaft as the spring is compressed. There must also be a hole cut in the side of the U-shaped device in order that a screwdriver blade or a magnetic screwdriver end can be passed through to enable collet removal. Place the complete unit, with the U-shaped piece in position, in a large vice or alternatively a large press, as might be found in a local builders or building merchants premises, and tighten down the vice/press. The method used in dismantling the project machine, as shown in photo 29.8p involved the use of a scarf clamp and an old car starter motor pinion. Whatever method is chosen, it need only be sufficient to compress the spring enough to allow the collets to be removed. With the collets displaced, release the tension on the spring slowly and remove the mainshaft assembly from the vice/press. Remove the spring top seat, slide off the spring and the lower internal spring guide. Great care should be taken to prevent the spring from escaping when being compressed.

7   The two sections of the cam-type damper unit can be removed and inspected. Inspect the dogs of the cam-piece, and the splined accepting cam-piece with which it mates, for wear of the cam profiles. If wear is pronounced, both components should be renewed. Examine the oil seal in the centre of the accepting cam-piece; good practice demands its renewal as a matter of course. The oil seal is retained in position by a circlip. Grease the new oil seal before installing it in the cam-piece. Similarly, displace the oil seal on the base of the accepting cam-piece, and replace it with a new item.

8   When reassembling the gear pinions on the outer mainshaft ensure that all the pinions are lubricated where they rotate on the shaft. Ensure all the oil holes in the spacer align with the holes drilled in the shaft. The circlip securing the 4th gear pinion must be fitted with the chamfered side towards the pinion. When installing the tabbed spacer between the 2nd gear and 5th gear drive pinions, ensure the tabs face away from the 2nd

gear pinion. Do not forget the right-hand washer which fits on the end of the outer mainshaft after the 2nd gear drive pinion has been fitted. Suzuki recommend that both sides of this washer should be treated, sparingly, with a coat of Suzuki Moly Paste (Part No. 99000-25140), before installation. With the gear pinions assembled on the outer mainshaft, apply, again sparingly, Suzuki Moly Paste to the area between the end of, and 20–30 mm (0.78 – 1.18 in) inwards along, the shaft. Liberally oil the bearing surfaces of the inner mainshafts prior to inserting the inner shaft into the outer shaft. When installing the dog-accepting cam-piece on the outer mainshaft, ensure the cut-out on the cam-piece inner face aligns with the oil hole in the shaft. Apply a further coat of Suzuki Moly Paste to the internal splines of the dogged cam-piece, prior to fitting it to the inner mainshaft. Place the spring internal guide on the inner shaft, followed by the large spring, and the top spring seat. The reverse of the collet removing procedure must now be followed, with the spring again being compressed, only this time to allow the collets to be refitted. Ensure the collets are firmly in position before releasing SLOWLY the pressure on the spring.

9   Give the gearbox components a close visual inspection for signs of wear or damage such as broken or chipped teeth, worn dogs, damaged or worn splines and bent selectors. Replace any parts found unserviceable because they cannot be reclaimed in a satisfactory manner.

10 The gearbox bearings must be free from play and show no signs of roughness when they are rotated. After thorough washing in petrol the bearings should be examined for roughness and play. Also check for pitting on the roller tracks.

11 It is advisable to renew the gearbox oil seals irrespective of their condition. Should a re-used oil seal fail at a later date, a considerable amount of work is involved to gain access to renew it.

12 Check the gear selector rods for straightness by rolling them on a sheet of plate glass. A bent rod will cause difficulty in selecting gears and will make the gear change particularly heavy.

13 The selector forks should be examined closely, to ensure that they are not bent or badly worn. The pegs which engage with the cam channels are integral with the forks therefore if they are worn the forks must be renewed. Under normal conditions, the gear selector mechanism is unlikely to wear quickly, unless the gearbox oil level has been allowed to become low.

14 The tracks in the selector drum, with which the selector forks engage, should not show any undue signs of wear unless neglect has led to under lubrication of the gearbox. Check the tension of the gearchange arm and drum stopper arm springs. Weakness in the springs will lead to imprecise gear selection. Check the condition of the gear stopper arm roller. It is unlikely that wear will take place here except after considerable mileage.

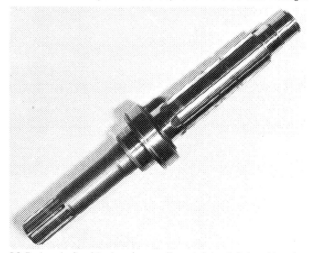

29.5a Layshaft with shrunk-on collar retaining left-hand bearing

29.5b Reassemble layshaft starting with 1st gear driven pinion (35T) and ...

29.5c ... secure it with a splined washer and circlip

29.5d Fit 4th gear driven pinion (27T) and secure it with ...

29.5e ... a circlip followed by ...

29.5f ... a splined washer

29.5g Fit bush, ensuring oil hole aligns with hole in shaft followed by ...

29.5h ... the third gear driven pinion (29T)

29.5i Fit the special dogged locking washer and the splined washer and ...

29.5j ... locate them correctly before ...

29.5k ... fitting the 5th gear driven pinion (25T) ensuring ...

29.5l ... that the oil holes line-up

29.5m Fit the 2nd gear driven pinion (32T) ...

29.5n ... then a large plain washer, and then ...

29.5o ... the right-hand bearing

29.5p Secure the bearing with the circlip

29.5q Grease oil seal and fit to left-hand end of layshaft

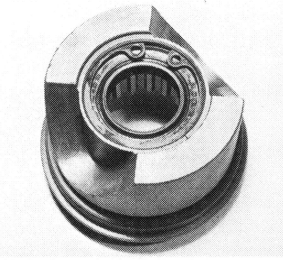

29.7a Examine cam piece for wear; note circlip securing central oil seal

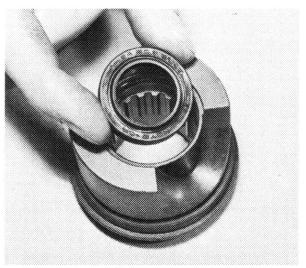

29.7b Replace central oil seal upon reassembly, and ...

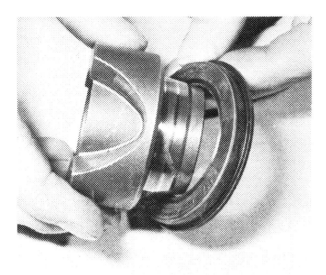

29.7c ... larger oil seal at base of cam piece

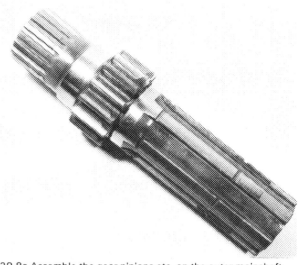

29.8a Assemble the gear pinions etc, on the outer mainshaft

29.8b Fit the sealed bearing to the left-hand end of the outer mainshaft

29.8c Ensure spacer oil hole aligns with hole in shaft and fit 4th gear drive pinion (24T)

29.8d Position splined washer and special dogged locking washer correctly

29.8e Fit the 3rd gear drive pinion (21T), followed by ...

29.8f ... a split ring and splined lock washer

29.8g Fit tabbed spacer to engage with notch in splined washer, ensure the oil holes align and ...

29.8h ... then fit the 5th gear drive pinion (26T)

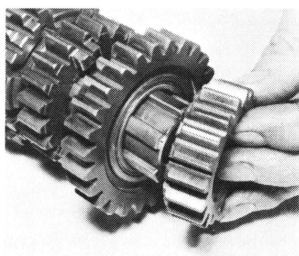

29.8i Fit the 2nd gear drive pinion (18T), followed by the end plain washer (not shown)

29.8j Engage the inner mainshaft with the outer shaft; note the left-hand bearing now becomes the middle bearing

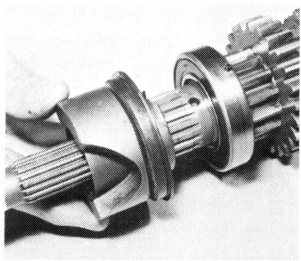

29.8k Fit the cam piece, then the ...

29.8l ... dogged cam-piece, greasing where applicable (see text)

29.8m Fit the spring internal guide and then ...

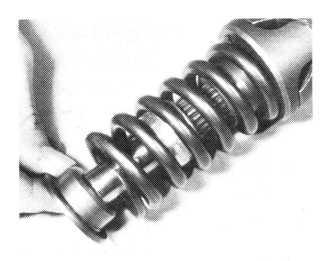

29.8n ... the spring and top seat

29.8o Apply grease to the collets before they are secured in place

29.8p Using the special tool and vice/press, refit the spring retaining collets

29.8q Ensure the collets are **secure** before releasing the clamping pressure

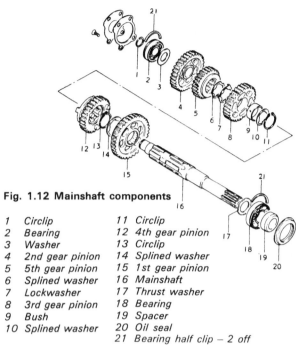

Fig. 1.12 Mainshaft components

1	Circlip	11	Circlip
2	Bearing	12	4th gear pinion
3	Washer	13	Circlip
4	2nd gear pinion	14	Splined washer
5	5th gear pinion	15	1st gear pinion
6	Splined washer	16	Mainshaft
7	Lockwasher	17	Thrust washer
8	3rd gear pinion	18	Bearing
9	Bush	19	Spacer
10	Splined washer	20	Oil seal
		21	Bearing half clip – 2 off

Fig. 1.14 Correct positioning of layshaft 3rd gear pinion retaining circlip

30 Examination and renovation: kickstart components – GS 850 GN model only

1 Due to the reliability of the electric start system, it is unlikely that the kickstart will be used sufficiently to induce wear. Breakage of the kickstart return spring, which is probably the most common fault, can be rectified by direct replacement of the damaged spring. The old spring may be easily removed from the shaft with a pair of pliers. When fitting a new spring, ensure that the inner turned end is located correctly in the radially drilled hole in the shaft.

2 Check the condition of the kickstart components. If slipping has been encountered a worn ratchet and pawl will invariably be traced as the cause. Any other damage or wear to the components will be self-evident. If either the ratchet or pawl is found to be faulty, both components must be replaced as a pair.

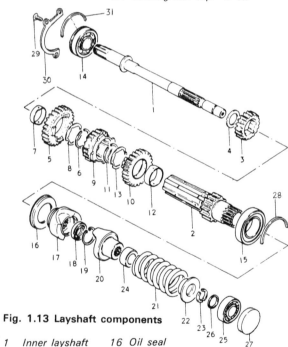

Fig. 1.13 Layshaft components

1	Inner layshaft	16	Oil seal
2	Outer layshaft	17	Output cam dog
3	2nd gear pinion	18	Oil seal
4	Thrust washer	19	Circlip
5	5th gear pinion	20	Input cam dog
6	Circlip	21	Spring
7	Spacer	22	Spring seat
8	Splined washer	23	Collets
9	3rd gear pinion	24	Spring guide
10	4th gear pinion	25	Bearing
11	Circlip	26	Spacer
12	Spacer	27	Plug
13	Splined washer	28	Bearing half clip
14	Bearing	29	Screw – 3 off
15	Bearing	30	Bearing retaining plate
		31	Bearing half clip

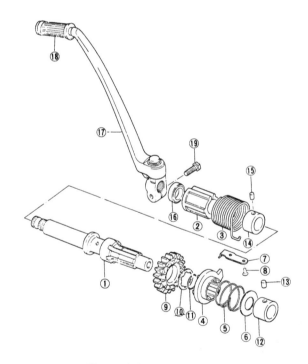

Fig. 1.15 Kickstart assembly

1	Kickstart shaft	11	Circlip
2	Spring guide	12	Bush
3	Spring	13	Pin
4	Ratchet pawl	14	Bush
5	Spring	15	Pin
6	Spring seating	16	Oil seal
7	Guide plate	17	Kickstart lever
8	Screw – 2 off	18	Lever rubber
9	Drive pinion	19	Bolt
10	Splined washer		

31 Examination and renovation: clutch assembly

1   After an extended period of use the clutch linings will wear and promote clutch slip. The clutch plates should be measured with a vernier gauge or pair of calipers to ascertain the extent of wear. The measurement of the thickness of the inserted (friction) plates and the maximum wear limit are as follows.

Inserted plate thickness   2.7 – 2.9 mm (0.106 – 0.144 in)
Wear limit   2.4 mm (0.094 in)

If the plate thickness is less than the specified minimum, then the plate must be renewed.

2   The clutch plates should not show any evidence of overheating (bluing). If they do, check them for overall flatness by placing each plate on a flat surface and measuring the bow with a feeler gauge. In the event of a plain plate being warped by more than 0.2 mm (0.008 in) it should be renewed.

3   A damper unit, consisting of a large wave washer and its seat, is fitted behind the final, innermost, plain clutch plate around the clutch centre boss. The wave washer, its seat and the last plain plate are all retained in position by a thin spring steel clip. If a pronounced chattering noise has been experienced from the clutch, or the engagement has become sharp, then the damping of the wave washer may be at fault. To renew the wave washer, displace the spring clip from the centre boss with a pair of pliers, and remove the innermost plain plate, followed by the washer. When reassembling the damper unit, the spring clip should be renewed.

4   Check the free length of each clutch spring. If the springs have shortened (set) to a length less than the specified minimum, they must be renewed. Renew the springs as a set, rather than individually.

Clutch spring free length   40.4 mm (1.59 in)
Wear limit   38.5 mm (1.52 in)

5   Check the condition of the clutch centre spacer and the large external caged needle roller bearing. If wear is evident in these components, they should be renewed. The bearing and spacer upon which the oil pump drive gear (fitted behind the clutch) is mounted should also be similarly checked.

6   Check the condition of the slots in the outer surface of the clutch centre and the inner surface of the outer drum. In any extreme case, clutch chatter may have caused the tongues of the inserted plates to make indentations in the slots of the outer drum, or the tongues of the plain plates to indent the slots of the clutch centre. These indentations will trap the clutch plates as they are freed and impair clutch action. If the damage is only slight, the indentations can be removed by careful work with a file and the burrs removed from the tongues of the clutch plates in similar fashion. More extensive damage will necessitate renewal of the parts concerned.

7   Check the clutch release thrust bearing and washer in the pressure plate. If play is evident in the bearing or if it rotates roughly, the bearing should be renewed.

8   The clutch release mechanism in the clutch cover does not normally require attention, provided that it is greased from time to time. If the unit fails, it must be renewed as a complete assembly. If any part requires renewal ensure that it is obtained from a Suzuki Service Agent; several modifications have been made, particularly to the release rack, to minimise clutch drag when cold.

32 Examination and renovation: secondary drive and driven bevel gear units

1   The two 90° bevel drive units which form the secondary drive and driven gear assemblies are self-contained separate units, detachable from the engine as complete sub-assemblies. Each unit is retained to the engine by four bolts passing through the flange on each housing. The secondary drive unit is fitted in the rear left-hand side of the engine, and the driven unit to the rear of the engine.

2   If excessive wear develops in the bevel gears of either unit, or if any of the internal components fail, the complete undismantled unit(s) should be returned to a reputable Suzuki service agent. It is strongly recommended that the bevel gear units be inspected and overhauled by qualified and experienced operators only, and for this reason no service or adjustment figures are given. It is an unfortunate fact that with this type of secondary drive mechanism, there are a large number of tasks that cannot be considered suitable for the home mechanic to attempt. This is, in part, due to the owner/rider being unlikely to possess the necessary skills, but more due to the necessary complications involved in setting up the bevel gears, and their associate components, to the exact tolerances required. This not only requires a high degree of skill and expertise, but would necessitate an extremely large outlay on specialised, single-purpose, service tools and service replacement and adjustment parts.

3   It is perhaps, pertinent at this stage, to note that due to the almost ideal conditions under which these bevel gear units operate – they are fully enclosed and well lubricated – wear is not a common occurence. It is widely known that this system of secondary transmission offers an extremely high degree of longevity with only limited maintenance. Excessive wear is only likely to occur if routine maintenance tasks such as topping up and changing the oil are neglected for a very long period.

4   Whenever the drive or driven units are removed from the engine, the O-rings attached to the housing of each unit should be detached and renewed before reassembly. When inserting the units in the engine, the O-rings should be lightly greased.

31.3a Innermost plain plate and damper retained by spring steel band

31.3b Displace spring retainer from centre boss to free innermost plate and damper

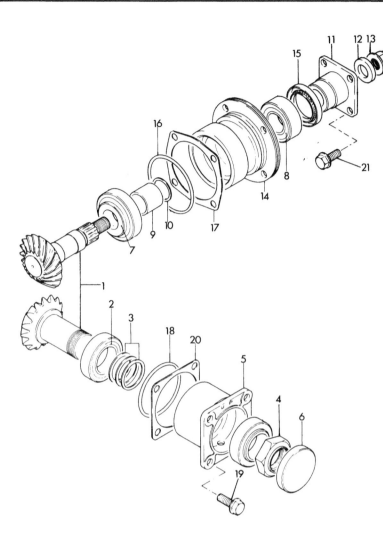

Fig. 1.16 Shaft drive unit

1   Secondary drive gears
2   Bearing
3   Shim
4   Nut
5   Input flange
6   Blanking plug
7   Bearing
8   Bearing
9   Spacer
10 Shim
11 Output flange
12 Washer
13 Nut
14 Housing
15 Oil seal
16 O-ring
17 Shim
18 O-ring
19 Bolt – 4 off
20 Shim
21 Bolt – 4 off

33 Crankcase covers: examination and renovation

1   The right-hand and left-hand crankcase covers and the inspection covers are unlikely to become damaged unless the machine is dropped or involved in an accident. Cracks in a casing can be repaired easily by special aluminium welding, provided the damage is not too extensive and care is taken to prevent distortion.

2   The covers are lightly polished and lacquered before leaving the factory. Badly scratched covers can be refurbished using a single cut file treated with chalk to prevent clogging, and finished off with fine emery paper and metal polish or aluminium cleaner. If required, the cases can be relacquered, using an aerosel paint spray.

34 Engine reassembly: general

1   Before reassembly of the engine/gearbox unit is commenced, the various component parts should be cleaned thoroughly and placed on a clean sheet of paper, close to the working area.

2   Make sure all traces of old gaskets have been removed and that the mating surfaces are clean and undamaged. One of the best ways to remove old gasket cement is to apply a rag soaked in methylated spirit. This acts as a solvent and will ensure that the cement is removed without resort to scraping and the consequent risk of damage. If a gasket becomes bonded to the surface through the effects of heat and age, a new sharp scalpel blade should be used to effect removal. Old gasket compound can also be removed using a soft brass wire brush of the type used for cleaning suede shoes. A considerable amount of scrubbing can take place without fear of damaging the mating surfaces.

3   Gather together all the necessary tools and have available an oil can filled with clean engine oil. Make sure that all new gaskets, oil seals and O-rings etc, are to hand, also all replacement parts required. Nothing is more frustrating than having to stop in the middle of a reassembly sequence because a vital gasket or replacement has been overlooked.

4   Make sure that the reassembly area is clean and that there is adequate working space. Refer to the torque and clearance settings wherever they are given. Many of the small bolts are easily sheared if overtightened. Always use the correct size screwdriver or bit for the crosshead screws, never an ordinary screwdriver or punch. If the existing screws show evidence of maltreatment in the past, it is advisable to renew them as a complete set.

5   In addition to the replacement parts, gasket cement etc., it will be necessary to obtain a bottle or tube of thread locking compound such as Loctite. This substance is used widely throughout the engine unit.

35 Engine reassembly: replacing the gear change drum and internal selector components

1   If the change drum stopper plate was removed from the drum for inspection or renewal, it must be refitted at this stage. The plate must be positioned so that it locates correctly with the drive pin which is a push fit in the drum end boss. Assemble the gear selector quadrant together with the two selector pawls, the plungers and the springs. The pawls must be fitted so that the narrower edges adjacent to the plunger recesses face towards the rear of the selector quadrant. Depress the pawls against the spring pressure and insert the completed selector quadrant into the end of the change drum.
2   Lubricate the change drum needle roller bearing with

engine oil, and slide the drum into place in the gearbox. Refit the neutral stopper plunger, detent spring and the bolt.
3   Slide the selector fork rods into position through the right-hand gearbox wall and refit the selector forks. The two rearmost selector forks, which share the same rod, are identical. Check that the forks are positioned the correct way round. Slide the change drum stopper arm onto the front selector rod. With the components correctly in place push both rods fully home and reconnect the stopper arm spring with the anchor point in the casing.
4   Coat the two countersunk crosshead screws which retain the change drum retainer plate, and the two further screws which retain the pawl plate, with locking fluid. Replace the two plates. Rotate the change drum until it is in the neutral position and the neutral stopper locates.

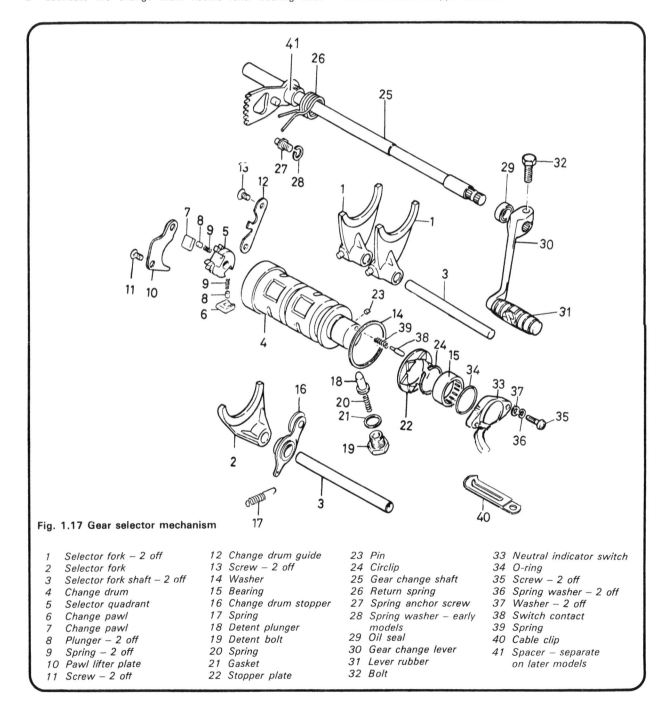

Fig. 1.17 Gear selector mechanism

1 Selector fork – 2 off	12 Change drum guide	23 Pin	33 Neutral indicator switch
2 Selector fork	13 Screw – 2 off	24 Circlip	34 O-ring
3 Selector fork shaft – 2 off	14 Washer	25 Gear change shaft	35 Screw – 2 off
4 Change drum	15 Bearing	26 Return spring	36 Spring washer – 2 off
5 Selector quadrant	16 Change drum stopper	27 Spring anchor screw	37 Washer – 2 off
6 Change pawl	17 Spring	28 Spring washer – early	38 Switch contact
7 Change pawl	18 Detent plunger	models	39 Spring
8 Plunger – 2 off	19 Detent bolt	29 Oil seal	40 Cable clip
9 Spring – 2 off	20 Spring	30 Gear change lever	41 Spacer – separate
10 Pawl lifter plate	21 Gasket	31 Lever rubber	on later models
11 Screw – 2 off	22 Stopper plate	32 Bolt	

35.2a Lubricate change drum needle roller bearing before fitting change drum

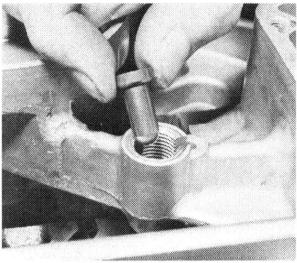

35.2b Fit the neutral detent plunger and ...

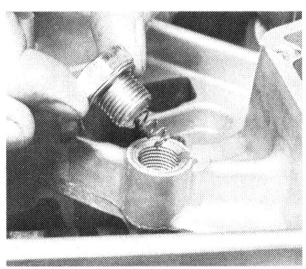

35.2c ... the detent housing and spring

35.3a Fit the forward selector rod and single fork; secure stopper arm spring as shown

35.3b Install rear selector rod and two selector forks

35.4 Apply locking fluid to the four countersunk screws

36 Engine reassembly: replacing the crankshaft

1 Position the upper crankcase half so that it rests on the cylinder holding down studs and the rear of the casing. Lubricate the main bearings thoroughly and also the crankshaft ends onto which are to be fitted the oil seals. The crankshaft right-hand seal should be fitted so that the cupped side is facing outwards, with the projections facing inwards in contact with the bearing. The left-hand seal is fitted with the spring garter side facing inwards.

2 Insert the five main bearing locating pins into the holes provided in the casing. The right-hand main bearing is located by a half clip in a manner similar to that of the gearbox ball bearings. In addition a small peg is fitted, which on installation of the crankshaft must be positioned in the recess adjacent to the bearing housing.

3 Fit the cam drive chain over the crankshaft so that it meshes with the drive sprocket. Grasp the crankshaft at both ends and lower the completed assembly into position. Ensure that the main bearing outer races engage correctly with the locating dowels. A tiny punch mark is provided on each main bearing outer race, diametrically opposed to the centre of the dowel hole. Lining up all five marks in a perfectly straight line

will aid correct location of the bearings. To aid the seating of each bearing firmly in its bearing housing, the gentle use of a soft-headed mallet is permissible. This should, of course, only be attempted after ensuring correct location of the locating pins and the half clip.

37 Engine reassembly: replacing the gear shaft assemblies

1 Position the upper crankcase half so that it rests on the cylinder holding down studs and the rear of the casing. Before refitting, the gearshafts must be assembled as completed sub-assemblies, including the gear pinions bearings and oil seals. The oil seal lips should be lubricated before being installed to prevent damage.

2 Install the bearing securing half clips in the casing grooves and lower the gear shafts into place, either individually or as a meshed pair. Note the location pins present on most of the bearings. Arrange these bearings so that the pins rests in the notches provided in the casing wall. Also locate the oil seal lips in their respective casing grooves.

3 Replace the oil seal/end plug at the blind end (left-hand end) of the mainshaft.

36.2 Install the five main bearing locating pins

36.3a Lower the crankshaft assembly into position

36.3b Dots on main bearing outer races aids alignment of pin holes

36.3c RH main bearing is located by half-clip (arrowed) and locating peg

36.3d Ensure right-hand end oil seal is fitted correctly as shown

37.2a Install the bearing securing half clips

37.2b Install the complete mainshaft/damper unit assembly ...

37.2c ... fit the LH bearing (note the locating groove in the casing outboard of the bearing) and ...

37.2d ... fit the oil seal/end plug with its flanged end filling the locating groove

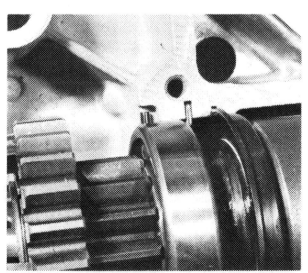
37.2e Arrange middle mainshaft bearing and oil seal as shown; locate peg, half-clip and oil seal lip

37.2f Install the complete layshaft, again ...

37.2g ... arranging the LH bearing locating peg, half-clip and oil seal as shown

38 Engine reassembly: replacing the kickstart assembly – GS 850 GN model only

1   If the kickstart shaft was dismantled for inspection, it must be reassembled before fitting into the crankcase. Refit the components into the kickstart shaft in the order shown in the accompanying illustration. Note that the ratchet pawl has a punch mark on the outer face must must be aligned with a similar mark on the shaft splines.

2   Ensure that the two small kickstart shaft locating pins are fitted correctly into the crankcase. Place the complete kickstart assembly into position.

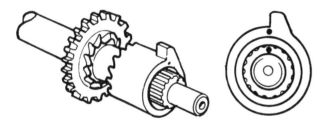

Fig. 1.18 Kickstart ratchet pawl punch marks

39 Engine reassembly: joining the crankcase halves

1   Carefully clean the crankcase halves mating surfaces. Replace the two hollow locating dowels, tapping them into position carefully so as not to distort them. If the dowels have become slightly burred they should be cleaned up with a small file.

2   Smear the upper crankcase half mating face with a thin layer of jointing compound. Suzuki recommend that Suzuki Bond No. 4 be used to make the joint. A good quality non-hardening compound will make a suitable substitute. Fit the small O-ring into the recess to the front of the mainshaft assembly.

3   Let the gasket compound set for at least 10 minutes and then lower the lower casing down into place. No difficulty should be encountered in fitting the upper case but special care should be taken that the three selector forks locate with their respective guide ways in the sliding pinions. To aid this operation, ensure that the change drum is in the neutral position

and arrange the gears so that they too are in neutral. When the lower casing is lowered into position, the selector forks fall naturally into a vertical position and so engage easily with the pinions.

4   Before refitting the crankcase bolts a check should be made on the operation of the reassembled gearbox components. Temporarily refit the gearchange lever and ensure that all five gears are engaging and disengaging in the correct manner. If the operation is not correct initially, the selector forks may not have engaged exactly with their respective guide ways. Care and patience may be required in order to manoeuvre the components into their respective positions. Remove the gearchange lever when the check has proved the operation of the gearbox to be correct.

5   Fit the crankcase retaining bolts to the lower crankcase. Tighten the thirteen 8 mm bolts evenly a little at a time, following the numerical sequence stamped on the bolts, commencing with No 1. Ensure the alternator lead guide clip is replaced on No 6 bolt. Fit and tighten, again evenly a little at a time, the eleven 6 mm bolts. Invert the crankcase, taking care not to let the cam chain become detached from its drive gear, and then fit the sixteen upper bolts. On early models there are four 8 mm bolts and twelve 6 mm bolts, while on later GT and GLT models there are six 8 mm bolts (four of which are secured by nuts, two of these having separate washers) and ten 6 mm bolts. All crankcase bolts should be tightened to the specified torque settings.

40 Engine reassembly: replacing the oil strainer screen, sump and oil filter

1   Place the oil strainer screen in position on the underside of the pick-up chamber and fit the retaining screws. A little locking fluid may be used on the screws, to ensure that they cannot become loose.

2   Fit a new sump gasket followed by the sump and retaining bolts. No gasket cement is required at this joint. Tighten the thirteen bolts evenly, to avoid distortion. Note that two of the bolts carry cable guide clips. Fit and tighten the oil drain plug, ensuring that the plug is not overtightened. With frequent recommended oil changes the thread can become worn. Too much force will strip the thread.

3   Insert a new oil filter element, with the O-ring end facing inwards, in the chamber at the front of the engine. Position the coil spring against the element and then refit the cover. Ensure the O-ring in the cover periphery is refitted; if it shows any signs of deterioration it should be renewed. Fit and tighten the three washers and domed retaining nuts.

38.2a Install the two locating pins in the casing and ...

38.2b ... fit assembled kickstart shaft

39.2 DO NOT omit this O-ring in crankcase upper half

40.1 Apply locking fluid to oil strainer screen screws

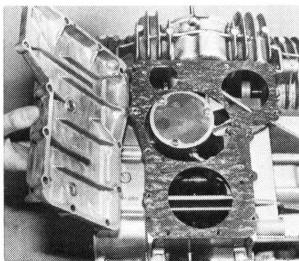

40.2a Fit a new sump gasket and sump plate

40.2b Install and tighten, carefully, the oil drain plug

41 Engine reassembly: replacing the gear selector external components

1   Refit the mainshaft bearing retainer plate in the primary drive/clutch housing and the layshaft blind end plate, and new gasket. These plates are retained by three and four countersunk screws respectively. In addition, if the lubrication passage end plate was detached for cleaning or inspection purposes, it should now be refitted. A new sealing gasket should be fitted prior to the plate being positioned and retained by three countersunk screws. Apply locking fluid to all these screws to ensure security.

2   Grease the gear change shaft oil seal on the left-hand side of the gearbox. Insert the change shaft, complete with the main change arm, into position, taking care that the splines on the shaft end do not damage the oil seal. A liberal coating of grease on the splines before the shaft is inserted into the casing, should ensure the splines of the shaft do not damage the oil seal as they pass through. Mesh the teeth on the change arm with the teeth on the change drum, as shown in the accompanying photograph. The change arm centraliser spring must be fitted

on the arm as shown. When in position the spring arms must lie either side of the anchor peg in the casing.

42 Engine reassembly: replacing the oil pump and oil pump drive gear

1   The oil pump must be re-installed in the primary drive casing as a complete unit, either before or after the driven gear is fitted. The driven gear is retained by a circlip on the shaft end, and is located by a drive pin which passes through the shaft, engaging with a recess in the rear face of the pinion.

2   Place a new O-ring in each of the casing recesses against the wall of which the pump is secured. Omission of the O-rings will lead to lubrication failure. Position the oil pump and fit and tighten the three mounting screws.

3   The oil pump drive gear assembly can now be replaced. Fit the heavy backing washer onto the clutch shaft, followed by the drive gear bearing spacer and needle roller bearing. Lubricate the bearing thoroughly and then refit the drive pinion. The pinion must be fitted with the two projecting dogs facing outwards, as it is these that locate with the rear of the clutch drum and so provide the driving medium.

41.1a Use a new sealing gasket when fitting lubrication passage end plate and ...

41.1b ... the layshaft blind end plate

41.1c Three further countersunk screws secure mainshaft bearing retainer plate

41.2 Install change shaft, mesh change arm with quadrant. Ensure change arm centraliser spring is correctly fitted (arrowed)

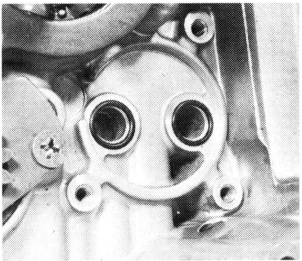

42.2a Fit the two new O-rings into the casing before ...

42.2b ... installing and securing the oil pump

42.2c Oil pump drive gear is located by a drive pin and ...

42.2d ... secured by a small circlip

42.3a Fit the heavy backing washer, followed by ...

42.3b ... the oil pump drive pinion and bearing. Note two projections must face outwards (arrowed)

43 Engine reassembly: replacing the clutch

1   Place the clutch outer drum over the clutch shaft (mainshaft) and into the casing. Centralise the drum and after lubrication, fit the spacer and needle bearing. The spacer should be fitted with the radially grooved face inwards. When fitting the outer drum, it is essential that the two projecting dogs on the oil pump drive gear engage with two recesses in the drum rear face. Insert a narrow shanked screwdriver through one of the two holes in the clutch spacer so that the drive pinion is prevented from rotating. Turn the clutch outer drum until it can be felt that the pinion and drum have engaged correctly.

2   Install the heavy thrust washer on the clutch shaft, with the milled channel facing inwards, and then refit the clutch centre boss onto the clutch shaft splines. Fit the tab washer and then fit and tighten the centre nut. Use the same procedure for tightening the nut as was used for loosening. Refer to Section 14, paragraph 3 of this Chapter. After tightening the nut, do not omit to bend up the tab washer to secure the nut in place.

3   The clutch plates should now be replaced one at a time. If the innermost plain plate was removed, to gain access to the wave washer damper or for examination purposes, it should be replaced and secured with a *new* spring steel retaining clip. If this plain plate is already in place, commence by replacing a friction (inserted) plate followed by a second plain plate, and so on alternately. Lubricate the clutch thrust piece and fit it, together with the thrust bearing and shim. Refit the clutch pressure plate and the clutch springs, washers and bolts. Tighten the bolts evenly, a little at a time, down to a torque setting of 1.1 – 1.3 kgf m (8.0 – 9.5 lbf ft).

44 Engine reassembly: replacing the kickstart return spring – GS 850 GN model only

1   Place the kickstart return spring over the kickstart shaft, with the outer turned end facing towards the gearbox wall. Insert the inner turned end of the spring in the axial drilling in the shaft. Rotate the shaft clockwise as far as possible so that it comes up against the stop. Using a stout pair of pliers, grasp the outer end of the spring and tension the spring in a clockwise direction until the turned end can be located in the anchor recess in the casing. With the spring tensioned correctly, insert the central spring guide.

45 Engine reassembly: replacing the clutch/primary drive cover

1   The primary drive cover can now be replaced. Lubricate the primary drive gears with engine oil and then fit a new gasket to the mating surface. Lightly grease the teeth on the clutch thrust piece (release rack) and the teeth on the pinion attached to the release shaft, fitted inside the cover. Whilst the grease can is to hand, liberally apply grease to the splines on the end of the kick start shaft. This will ensure the splines do not damage the oil seal in the outer cover as it is refitted.

2   Push the cover into position on the two hollow locating dowels, tapping gently with a soft-headed mallet if necessary, and fit the ten retaining screws. The screws should be tightened evenly, in a diagonal sequence, to prevent distortion. Care should be taken to ensure that the teeth on the clutch release thrust piece (release rack) engage with those on the release shaft pinion.

46 Engine reassembly: replacing the starter motor clutch and intermediate gear and alternator

1   Place the starter clutch pinion bearing thrust washer onto the left-hand end of the crankshaft so that the face with the chamfered inner radius is towards the main bearing. Lubricate the double needle roller bearings and fit them, together with the clutch pinion, if it was separated from the body of the starter clutch.

2   Before fitting the combined alternator rotor/starter clutch unit, clean thoroughly the external taper on the crankshaft end and the internal taper in the alternator rotor. Position the rotor/clutch assembly on the shaft, turning it anti-clockwise so that the three clutch rollers slide easily onto the pinion boss. Insert and tighten the rotor centre bolt The correct torque wrench setting is 9.0 – 10.0 kgf m (65.0 – 72.5 lbf ft).

3   Place the intermediate double gear on its stub shaft and fit one thrust washer either side of the gear. Position the assembly in the casing with the larger gear to the rear of the starter clutch pinion, and push the stub shaft fully home into the casing recess.

4   Fit a new gasket to the crankcase surface and place the alternator cover close to the engine so that the alternator leads may be threaded through the wall of the starter clutch chamber into the starter motor compartment. Pull the wires through whilst fitting the outer cover. Fit and tighten evenly the casing holding screws. Gentle persuasion from a soft-headed mallet may be required to fully seat the casing on its single locating dowel.

43.1 Install the clutch outer drum

43.2a Heavy washer must be fitted, plain face outwards

43.2b Fit clutch centre boss, tab washer and centre nut ...

43.2c ... tightening the nut to the correct torque setting. Note use of locking tool to prevent clutch rotation

43.3a Replace plain and friction plates alternately, one at a time

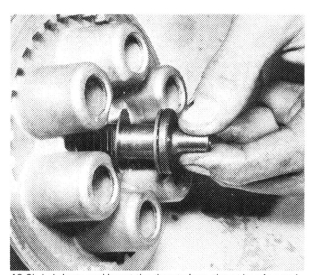

43.3b Lubricate and insert the thrust piece, thrust bearing and washer

43.3c Refit the clutch pressure plate

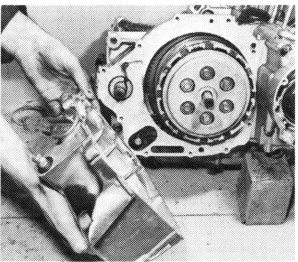

45.1 Using a new gasket, refit the primary drive/clutch cover

46.1 Fit starter clutch pinion bearing thrust washer correctly as shown

46.2 Lubricate the needle roller races and fit the clutch pinion/starter clutch

46.3 Fit thrust washer each side of intermediate gear

46.4 Feed the alternator wires through the casing

47 Engine reassembly: replacing the secondary drive and driven bevel gear units

1   As has already been stated, the two bevel drive units which form the secondary drive train, are complete separate sub-assemblies and may be refitted as such. Each unit is secured to the engine by four bolts passing through the flange attached to each housing.

2   Before fitting either unit, ensure that the metal shims attached to each housing are **exactly** as they were upon removal. This does not, of course, apply in the case of a new or refurbished unit. In this case the shimming, and other associate setting-up operations. **must** be carried out by a Service Agent at the time of fitting the units to the engine. Ensure that each bevel gear unit is completely clean, and lubricate the teeth fully with clean hypoid (SAE 90) gear oil.

3   Install a new O-ring on each of the housings and lightly grease them prior to insertion into the engine. At this juncture the help of an assistant may well prove beneficial. Refit the gear change lever and engage 2nd gear or higher ratio. Instruct your assistant to grasp the cam chain firmly where it passes up through the crankcase, and, when you indicate to him to do so, to pull on the cam chain. This action will rotate the crankshaft which will in turn rotate all the drive train shafts and ease engagement for the teeth on the bevel drive units.

4   Insert the secondary drive unit to the left-hand side of the engine. Rotate the engine as described above to ease the meshing operation. With the teeth on the bevel gear unit correctly engaged, gentle persuasion with a mallet and block of wood, to assist complete seating of the drive unit in the casing, is permissible. Ensure that the UP mark on the housing flange is indeed at the top of the unit when installed. Coat the four retaining bolts with locking fluid and tighten them down to a torque wrench setting of 2.0 – 2.6 kgf m (14.5 – 19.0 lbf ft).

5   Refit the secondary driven bevel gear unit to the rear of the engine. Lubricate the housing and its bore with clean hypoid (SAE 90) gear oil to ease fitting. Rotate the engine in the same way as for the drive unit, to aid the bevel gear meshing operation. Ensure the UP mark on the housing flange is correctly positioned. Apply a coat of locking fluid to the four retaining bolts, fit them to the housing flange, and tighten them down to a torque wrench setting of 2.0 – 2.6 kgf m (14.5 – 19.0 lbf ft).

6   Install the secondary gearbox drain plug and tighten the plug to a torque setting of 2.0 – 3.0 kgf m (14.5 – 21.5 lbf ft).

47.4 With the secondary drive unit fitted and secured, install the secondary driven unit to the rear

47.5 Ensure the UP mark on the housing flange is at the top when installed

48 Engine reassembly: replacing the neutral indicator/gear position indicator switch

1   Position the O-ring in the switch body recess in the left-hand side of the casing, below and inset from the secondary drive gearbox. Insert the contact spring and the contact brush itself in the hole in the change drum end. Replace the neutral indicator/gear position indicator body, and insert the two retaining screws, noting the plain and lock washer on each screw.

49 Engine reassembly: replacing the ATU and contact breaker assembly – GS 850 GN model only

1   Position the ATU against the end of the crankshaft so that the drive pin projecting from the shaft end engages with the recess in the rear of the unit. Fit the timing index pointed plate into the casing so that scribed mark is at the 12 o'clock position, and then install the complete contact breaker assembly stator plate. The stator plate should be fittd so that the wiring lead grommet can locate with the rebated hole in the casing wall.
2   If, on dismantling, a punch mark or scribe mark was made on the stator plate aligning with a similar mark on the casing, these marks should be realigned and the three retaining screws fitted. It is probable that the ignition timing will be correct but a check must be made as a precautionary measure. Where no alignment marks were made, ignition timing should be set as a matter of course, as described in Chapter 3, Section 7.
3   Before checking or setting the ignition timing, replace the ATU centre bolt and the hexagonal engine turning piece. To tighten the bolt, pass a close fitting bar through one small-end eye, bearing down on two wooden blocks placed across the crankcase mouth. Do not fit the contact breaker cover before fitting and timing the valves

50 Engine reassembly: replacing the ATU and ignition signal generating unit – GS 850 GT/GLT models

1   Position the ATU against the end of the crankshaft so that the drive pin projecting from the shaft end engages with the

recess in the rear of the unit. Fit the timing index pointer plate into the casing so that the scribed mark is at the 12 o'clock position. The complete signal generating assembly on its mounting plate can then be installed in the casing. The mounting plate should be fitted so that the wiring lead grommet can locate with the rebated hole in the casing wall.
2   If, on dismantling, a punch or scratch mark was made on the mounting plate aligning with a similar mark on the casing, these marks should be realigned and the three retaining screws refitted. If no marks were made position the base plate so that it is approximately central to the mounting screws. In either case the ignition timing must be checked as described in Chapter 3, Section 7.
3   Replace the ATU centre bolt and the hexagonal engine turning piece. To tighten the bolt, adopt the same procedure described in paragraph 3 of the previous Section.

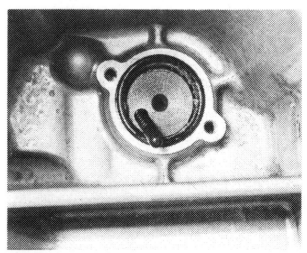

48.1a Fit the contact brush and spring

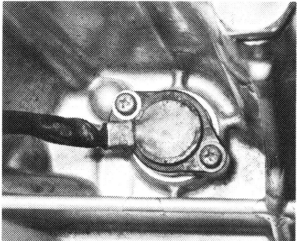

48.1b Refit switch body, not forgetting the O-ring

49.1a Install ATU so that its notch locates with drive pin

49.1b Fit index plate in position shown

49.1c Refit stator plate assembly in original position; feed wiring lead through casing grommet

49.3 Fit ATU centre bolt and engine turning hexagon

51 Engine reassembly: replacing the pistons and the cylinder block

1   Fit the piston rings to each piston, commencing with the three-piece oil control ring. The first of the three parts to be fitted should be the corrugated spacer band. Fit the oil control ring side rails one at a time. When fitting the two compression rings, ensure that both are fitted with the R mark facing upwards. Do not mix rings of different makes in the same engine. The two compression rings are of differing type and cross-section. The upper ring has a chrome plated face which is slightly curved. The 2nd ring is not chrome plated and has a tapered face. Before replacing the pistons, pad the mouths of the crankcase with rag in order to prevent any displaced component from accidentally dropping into the crankcase.
2   Fit the pistons in their original order with the arrow on the piston crown pointing toward the front of the engine.
3   If the gudgeon pins are a tight fit, first warm the pistons to expand the metal. Oil the gudgeon pins and small-end bearing surfaces, also the piston bosses, before fitting the pistons.
4   Always use new circlips **never** the originals. Always check that the circlips are located properly in their grooves in the piston boss. A displaced circlip will cause severe damage to the cylinder bore, and possibly an engine seizure.

5   Replace the chain tensioner rear guide blade, ensuring that the lower end locates correctly with the guide seat retained in the crankcase. Fit and tighten the single securing bolt. Install the two sealing rings in the recesses surrounding the outer rear cylinder studs and then fit a new cylinder base gasket over the studs. Check that the two hollow locating dowels are fitted to the outer front holding down studs. A new O-ring should be placed on each of the four cylinder bore spigots. Push the O-rings fully home, so that they seat correctly in the grooves provided.

6   Arrange the piston ring gaps as shown in the accompanying diagram, in order to maintain the best sealing characteristics. Using clean engine oil, lubricate thoroughly the cylinder bores. Lift the cylinder block up onto the studs and support it there whilst the camshaft chain is threaded through the tunnel between the bores. The task is best achieved by using a piece of stiff wire to hook the chain through, and pull up through the tunnel. The chain must engage with the crankshaft drive sprocket.

7   The cylinder bores have a generous lead in for the pistons at the bottom, and although it is an advantage on a large engine such as this to use the special Suzuki ring compressor, in the absence of this, it is possible to lead gently the pistons into the bores, working across from one side. Great care has to be taken **not** to put too much pressure on the fitted piston rings. When the pistons have finally engaged, remove the rag padding from the crankcase mouths and lower the cylinder block still further until it seats firmly on the base gasket.

8   Take care to anchor the camshaft chain throughout this operation to save the chain dropping down into the crankcase.

51.2 Pistons must be fitted with arrow mark facing forward

51.3 Push in gudgeon pin to secure piston

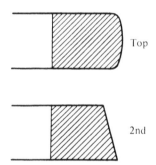

Fig. 1.19 Cross-section of top and second piston rings

51.4 Ensure that the **new** circlips are correctly fitted in their grooves

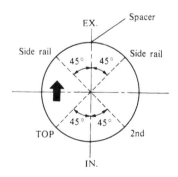

Fig. 1.20 Piston ring gap positions

51.5a Refit the chain tensioner blade

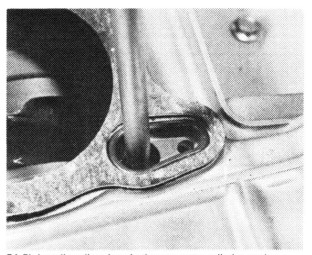

51.5b Install sealing rings in the rear outer cylinder studs recesses

51.7a Guide cylinder block over cam chain and ...

51.7b ... two at a time, ease the pistons into the bores taking care not to trap the rings

52 Engine reassembly: replacing the cylinder head

1   Place a new cylinder head gasket in position on the holding down studs. Fit the gasket so that the widest turned edges of the cylinder bore reinforcement inserts are facing downwards. Replace the cam tunnel rectangular seal.

2   Slide the cylinder head down the studs into position whilst guiding the cam chain through the tunnel. Secure the chain once again. Fit the cylinder head holding nuts, washers and bolts. It should be noted that the four outer studs are fitted with copper washers, to prevent oil seepage from the stud holes which serve also as oilways.

3   Tighten the cylinder head nuts evenly, in the sequence shown in the accompanying illustration, to a setting of 3.7 kgf m (27.0 lbf ft). Then, fit and tighten the two 6 mm bolts to a setting of 0.9 kgf m (6.5 lbf ft)/ Finally, insert and tighten the single forward 6 mm bolt, noting that it fits into the cam chain tunnel casting of the block and passes upwards into a lug projecting from the cylinder head. Tighten the 6 mm bolts to a setting 0.9 kgf m (6.5 lbf ft).

4   Insert the cam chain tensioner forward blade into the cylinder head central tunnel so that the lower end engages in the recess in the guide seat and the upper end locates with the

recesses each side of the tunnel. Make a special check that the lower end of the blade is restrained; if it is allowed to float in operation the cam chain may become damaged.

53 Engine assembly: replacing the camshafts and timing the valves

1   Lubricate thoroughly and insert the cam followers into their original guide tunnels in the cylinder head. Provided the followers are fitted squarely, very little pressure should be required to insert them in the tunnels. Do not attempt to tap (even lightly) a follower into place as it will probably jam solidly. Removal is then very difficult. Fit also the original adjuster shims. Lubricate the camshaft journals with a molybdenum disulphide grease and apply engine oil to the camshaft journal bearing surfaces.

2   Because the cam chain is endless, the camshaft replacement and valve timing operations must be made simultaneously. Fit a spanner to the engine turning hexagon and rotate the engine forwards until the T mark on the left of the F1-4 mark on the ATU is in **exact** alignment with the index mark on the static index plate. In this position Nos 1 and 4

cylinders are at TDC. Whilst turning the engine, the cam chain will have to be hand fed so that it does not become snagged or bunched up.

3 Select the exhaust camshaft (marked EX) and feed it through the cam chain. Pull the forward run of the chain taut and mesh the cam sprocket to the chain so that when the camshaft is lying across the bearing housings, the '1' marked arrow is lying horizontally in line with the cylinder head top mating surface. Fit the exhaust camshaft bearing caps, as detailed in paragraph 4, and then recheck the camshaft position. Insert the inlet camshaft through the camchain in a manner similar to that for the exhaust camshaft. To mesh the sprocket in the correct place, count the chain roller pins from the exhaust camshaft to the inlet camshaft. Start with the pin directly above the '2' marked arrow on the exhaust sprocket and count to the 20th pin along the chain. Mesh the '3' marked arrow on the inlet sprocket with the 20th pin. Provided that the crankshaft has not moved during this procedure, the valve timing is now correct. As a final check that the camshafts are in the correct position, the notch on the right-hand end of each camshaft should be horizontal, with the two notches facing each other.

4 The two bearing caps holding each camshaft should be refitted as follows after lubricating the journals thoroughly. Note that each cap is marked A, B, C or D and each should be fitted to the bearing housing similarly marked. The caps should be placed so that the letters are not inverted, relative to one another. The same problem is now encountered as was found on dismantling, in that the upward pressure of the valve spring is preventing immediate seating of the camshafts. The system of clamping the camshaft down, using a self-grip wrench or G-clamp may be adopted. Alternatively the bearing caps and bolts may be fitted and tightened down evenly and diagonally a small amount at a time. The latter procedure should be used only if great care is exercised and neither the camshaft nor the caps are allowed to tilt. Whichever procedure is chosen, the final tightening torque for the bearing cap bolts is 1.0 kgf m (7.0 lbf ft). After tightening the cap bolts, re-check the valve timing.

5 Fit the cam chain tensioner jockey sprocket. Ensure that the sprocket is fitted in its original position and tighten the four holding bolts to 0.6 – 0.8 kgf m (4.5 – 6.0 lbf ft). The performance of the rubber damper blocks upon which is mounted the sprocket depends upon the bolts being tightened to the correct torque.

6 Insert the tachometer driveshaft and housing into the front of the cylinder head and fit the retaining plate and bolt.

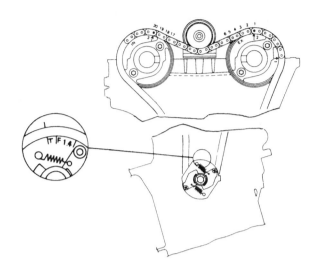

Fig. 1.22 Valve timing index marks

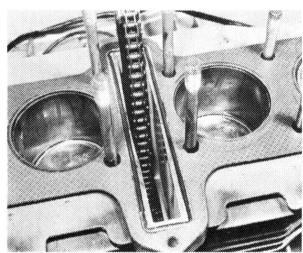

52.1 Fit a new cylinder head gasket and cam tunnel seal

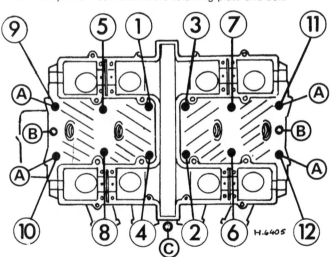

A = Domed nuts fitted with copper washers
B = Two 6 mm bolts
C = One 6 mm bolt

Fig. 1.21 Cylinder head nut tightening sequence

52.2 Ensure copper washer is fitted below each domed nut, note tightening sequence number stamped in head

52.3 Do not omit 6 mm bolt at front of cam chain tunnel casting

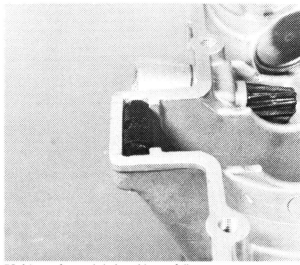

52.4 Insert forward chain guide carefully

53.3a Each camshaft is marked IN or EX and R or L to avoid confusion

53.3b Exhaust camshaft sprocket timing mark '2' should be vertical and '1' horizontal

53.3c Inlet cam sprocket timing mark '3' must be below the 20th pin

53.3d As a final check, note the position of the camshaft end notches (arrowed)

53.4 Fit cam bearing caps so that letters match with letters inscribed in head

54 Engine reassembly: replacing and adjusting the cam chain tensioner

1   Hold the cam chain tensioner in one hand and whilst restraining the plunger rod, slacken the locking screw a few turns. Push the plunger inwards, fully, simultaneously rotating the knurled adjuster wheel anti-clockwise. Continue turning until the plunger is fully retracted and the knurled adjuster has moved as far as possible. Tighten the lock screw to secure the pushrod.

2   Fit a new gasket to the tensioner body flange and install the completed unit to the rear of the cylinder block. Tighten the securing bolts evenly, to a torque setting of 1.0 kgf m (7.2 lbf ft).

3   With the tensioner fitted to the engine, unscrew the locking screw $\frac{1}{4} - \frac{1}{2}$ a turn so that the plunger is free to move forwards under tension from the spring. Without allowing further rotation of the locking screw, tighten the locknut to secure the screw.

4   The cam chain tensioner is now set for automatic adjustment in service. To check that the unit is functioning correctly, rotate the engine backwards to take up all the slack in the rear run of the chain. Whilst turning the engine backwards rotate the knurled adjuster wheel slowly as far as possible. Now turn the engine in a forward direction, which will have the effect of slackening the chain on the rear run. The knurled adjuster wheel should be seen to rotate in a clockwise direction as the plunger moves out and automatically tensions the chain.

5   **WARNING**. After initial adjustment of the cam chain tensioner, the tensioner will continue to function automatically. **DO NOT** under any circumstances rotate the knurled adjuster wheel either clockwise or anti-clockwise except when making this adjustment in the prescribed manner. Rotation of the wheel except at this stage will cause excessive chain tightness and will lead to tensioner damage and chain damage.

55 Engine reassembly: checking and adjusting the valve clearances

1   The clearance between each cam and cam follower must be checked and if necessary adjusted by removal of the existing adjuster pad and replacement of a pad of suitable thickness. Make the clearance check and adjustment of each valve in sequence and then go on to the next valve. As shown in the diagram in the Routine Maintenance section both operations should be carried out with the cam lobe in question placed in one of two alternative positions.

2   Using a feeler gauge, verify and record the clearance at the first valve. If the clearance is incorrect, not being within the range 0.03 – 0.08 mm (0.001 – 0.003 inch), the adjuster pad must be removed and replaced by a shim of suitable thickness. A special tool is available (Suzuki part no. 09916-64510) which may be pushed between the camshaft adjacent to the cam lobe and the raised edge of the cam follower, to allow removal of the shim. If the special tool is not available, a simple substitute may be fabricated from steel plate. The final form of the tool which has a handle approximately 6 inches long is shown in the photograph accompanying the cam adjustment section in the Routine Maintenance Chapter. The Suzuki tool may be pushed into position, depressing and securing the cam follower in a depressed position in one operation. Where a home-made tool is used, the cam follower may be depressed by inserting a suitable lever between the adjuster pad and the cam lobe. The tool may then be inserted to secure the cam follower whilst the adjuster pad is removed. Before installing either type of tool, rotate the cam follower so that the slot in the follower raised edge is not obscured by the camshaft. Insert a screwdriver through the slot to displace the adjuster pad.

3   Adjustment pads are available in 20 sizes ranging from 2.15 mm to 3.10 mm, in increments of 0.05 mm. Each pad is identified by a three digit number etched on the reverse face. The number (eg 235) indicates that the pad thickness is 2.35 mm thick. To select the correct pad subtract 0.03 mm from the measured clearance and add the resultant figure to that of the existing pad. Select the largest available pad whose thickness is slightly smaller than the final figure. Refer to the table accompanying the Routine Maintenance Section for the selection of available pads. Although the adjustment pads are available as a set, their price is prohibitive. It is suggested that pads are purchased individually, after an accurate assessment of requirements has been made. It is possible that some Suzuki Service Agents will be prepared to exchange used pads for others of the correct size, provided that the original pads are not worn.

4   Before inserting a replacement pad, lubricate both sides thoroughly with engine oil. Always fit the pads with the number downwards, so that it does not become obliterated by the action of the cam. Recheck the valve clearance after fitting the new adjuster pads.

5   After restoring the clearances on each valve refit the camshaft cover. Do not omit the semi-circular seals which are located at each end of the camshaft chambers. Replace the camshaft cover end caps.

54.2 Fit chain tensioner, using a new gasket

54.5a Install the cam box end seals ...

54.5c ... refit the camshaft cover

56 Engine reassembly: replacing the engine in the frame

1   The task of replacing the engine requires three people, two to lift the engine and one to hold the frame steady while the engine is lowered into position.

2   Lift the complete engine unit into the frame from the right-hand side of the machine, and manoeuvre the rear of the engine over so that the final driveshaft gaiter can be pulled across to the right and hitched against the output flange. The front of the engine can be moved across so that the output flange (attached to the secondary driven unit) and driveshaft flange can align.

3   Mount but do not secure the four engine mounting brackets, before inserting the engine bolts. Insert the three long bolts from the left-hand side of the machine and fit the two short central bolts and the special plate bolts. Also insert the two further, shorter, bolts to the small mounting bracket on the lower right-hand side of the engine. Do not force the bolts into position as this may damage the threads. Use a wooden lever between the frame and the casings, in order to lift the engine and align bolt holes. Tighten the engine mounting bracket bolts first and then the engine bolts. The longer, 10 mm engine mounting bolts should be tightened to a torque setting of 3.5 kgf m (25.5 lb ft), and the 8 mm bolts to a torque setting of 2.5 kgf m (18.0 lb ft).

54.5b ... fit a new gasket and ...

4   Replace the four output flange/final driveshaft flange securing bolts, turning the flanges as necessary to gain access. The securing bolts should be treated to a coat of locking compound before tightening evenly to a torque setting of 2.5 – 3.0 kgf m (18.0 – 21.5 lbf ft). Relocate the rubber gaiter over the flange joint and secure it with the large screw clip.

5   Replace the starter motor in the compartment and fit and tighten the two securing bolts which pass through the inner end flange. Reconnect the starter motor lead with the terminal on the motor body and refit the rubber boot. Replace the chromed cover and secure with the two bolts. Remake the connections from the alternator, contact breakers or ignition signal generating unit, neutral/gear position indicator switch, and oil pressure warning lamp leads. Ensure that the leads are routed correctly and secured as necessary. Ensure that the colour coding of the wires is followed exactly. Reconnect the main earth lead to the crankshaft bolt which secures it. Note that the lead from the neutral/gear position indicator switch should be routed to pass below the secondary drive unit cover.

6   Replace the secondary drive unit cover, and secure it with the four screws.

7   Reconnect the clutch cable at the handlebar lever, if it was detached here, and then refit the chromed release arm to the operating shaft. Secure the release arm with the pinch bolt. Adjust the cable by means of the adjuster and locknut on the lower end of the cable, so that there is 2 – 3 mm (0.08 – 0.12 in) play measured between the stock and the handlebar lever, before the clutch commences lifting. Ensure the locknut is tightened after adjustment is correct.

8   Replace the breather cover on the cam box cover together with a new gasket, and then reconnect the HT leads with the sparking plugs, securing the leads below the guide clips provided. Each of the HT leads is marked by a numbered sleeve, to aid correct positioning.

9   Lift the carburettors into place from the right-hand side of the machine. Ensure that the screw clips are fitted to the induction stubs and air hoses before installing the carburettors. On the GS 850 GN model, refit the two throttle cables to the wheel type control lever (one cable opens the throttles, and one cable closes them), making sure that the opening cable is fitted to the rear, and the closing cable is fitted to the front of the operating wheel. When installing the closing cable, the tension in the strong return spring has to be pulled against; ensure it is not released suddenly or inadvertently. Reconnect the choke operating cable.

10   Position the air filter box to the rear of the carburettors, fit and tighten the two securing bolts on the top of the box, and then tighten all the screw clips fully.

11   Install the connecting hose between the breather cover and the air filter box. Secure the hose by means of the spring clips. On the GN model, the carburettor drain hoses should be

gathered together, and retained by a strap to the engine right-hand upper rear mounting bracket. Allow the trailing ends of the hoses to pass between the swinging arm member and the rear of the engine. The air vent pipes from carburettors No 2 and No 4 should be passed through the clamp attached to the rear of the air filter box. On the GT/GLT models, the single drain tube should be positioned and secured to the rear of the air filter box.

12 Refit the exhaust pipe and also the silencers if they too were removed. Tighten the exhaust pipe flange bolts to a torque setting of 0.9 – 1.4 kgf m (6.5 – 10.0 lbf ft), and the silencer mounting bolts to 1.8 – 2.8 kgf m (13.0 – 20.0 lbf ft). Refit the kickstart lever (GN model only) and the gear change lever. Check that both controls are in the correct operating positions before tightening the pinch bolts.

13 Replace the rider's footrests, tightening the securing bolts to a torque setting of 2.7 – 4.3 kgf m (19.5 – 31.0 lbf ft).

14 If the two horns were removed previously, they must now be replaced. When refitting, ensure that they are positioned correctly and will not come into contact with the petrol tank during the running of the machine. Remake the connections at the two snap connectors on each horn.

15 Lower the petrol tank into place and refit the rear retaining bolt and rubber seat. Connect the petrol pipe and the vacuum tube to the petrol tap unions. Secure the petrol pipe by means of the spring clip, and reconnect the two fuel gauge leads at their snap connectors. Reconnect the tachometer cable with its drive unit.

16 Check that the crankcase drain plug and the secondary drive gearbox drain plug are in place and tight. Refill the main engine/transmission lubrication system by pouring the specified quantity of oil into the primary drive/clutch casing. With the engine/transmission unit dry, after a complete stripdown and reassembly sequence, 3.8 litres (8.0/6.6 US/Imp pint) of SAE 10W/40 engine oil will be necessary to replenish the engine. The oil level can be checked through the sight window in the clutch cover which should be between the two marks. If, as the oil settles the level drops, more oil should be added as necessary.

17 Replenish the secondary drive gearbox with 340-400 cc (11.5 – 13.5/12.0 – 14.1 US Imp fl oz) of Hypoid SAE 90 gear oil. The oil should be poured through the filler orifice at the top of the secondary drive gearbox. Before the replenishment takes place, the oil level screw at the rear of the gearbox should be removed. When oil starts to flow from the oil level hole, the required amount of oil has been poured in. Refit the oil level screw and the filler plug.

18 Note that after the engine has been allowed to run for approximately 3 minutes after the inital start-up, the oil level should be rechecked. Ensure that the machine is standing vertically when checking both the oil levels because any angle of lean has a marked effect on the indicated levels.

19 Replace the battery and remake the connections. Give the terminals a coat of petroleum jelly to inhibit corrosion. Fit the frame side covers.

56.5 Insert the starter motor into its compartment

56.6 Replace and secure secondary drive unit cover

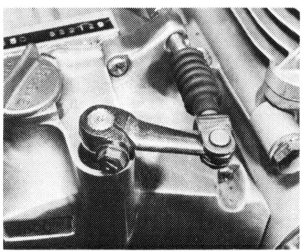
56.7 Refit clutch cable and release arm; adjust cable correctly

56.8 Replace breather cover and secure HT leads beneath clips

56.11 Secure air vent pipes as shown (GS 850 GN model)

56.12 Fit new exhaust port gaskets before refitting exhaust pipes

56.15 Secure tachometer cable with the knurled ring

56.16 Refill the engine with SAE 10W/40; check O-ring in filler cap

56.17 Replenish secondary drive gearbox with correct quantity of SAE 90 hypoid gear oil

57 Starting and running the rebuilt engine

1   Place the petrol tap in the 'Prime' position to allow the unrestricted flow of fuel to the carburettors. As soon as the engine is running revert to the normal 'On' position so that the induction vacuum will control the petrol flow. Close the carburettor chokes and start the engine using either the kickstarter (where fitted) or the electric starter. Raise the chokes as soon as the engine will run evenly and keep it running at a low speed for a few minutes to allow oil pressure to build up and the oil to circulate. If the red oil pressure indicator lamp is not extinguished, stop the engine immediately and investigate the lack of oil pressure.
2   The engine may tend to smoke through the exhausts initially, due to the amount of oil used when assembling the various components. The excess of oil should gradually burn away as the engine settles down.
3   Check the exterior of the machine for oil leaks or blowing gaskets. Make sure that each gear engages correctly and that all the controls function effectively, particularly the brakes. This is an essential last check before taking the machine on the road.

58 Taking the rebuilt machine on the road

1   Any rebuilt machine will need time to settle down, even if parts have been replaced in their original order. For this reason it is advisable to treat the machine gently for the first few miles to ensure that oil has circulated throughout the lubrication system and that any new parts fitted have begun to bed down.
2   Even greater care is necessary if the engine has been rebored or if a new crankshaft has been fitted. In the case of a rebore, the engine will have to be run-in again, as if the machine were new. This means greater use of the gearbox and a restraining hand on the throttle unit until at least 500 miles have been covered. There is no point in keeping to any set speed limit; the main requirement is to keep a light loading on the engine and to gradually work up performance until the 500 mile mark is reached. These recommendations can be lessened to an extent when only a new crankshaft is fitted. Experience is the best guide since it is easy to tell when an engine is running freely.
3   If at any time a lubrication failure is suspected stop the engine immediately, and investigate the cause. If an engine is run without oil, even for a short period, irreparable engine damage is inevitable.

59 Fault diagnosis: Engine

Symptom	Cause	Remedy
Engine will not start	Defective sparking plugs	Remove the plugs and lay them on the cylinder head. Check whether spark occurs when ignition is on and engine rotated.
	Dirty or closed contact breaker points (GN model only)	Check the condition of the points and whether the points gap is correct.
	Faulty or disconnected condenser (GN model only)	Check whether the points arc when separated. Renew the condenser if there is evidence of arcing.
	Faulty signal generating unit (GT/GLT models only)	See Chapter 3
Engine runs unevenly	Ignition or fuel system fault	Check each system independently, as though engine will not start.
	Blowing cylinder head gasket	Leak should be evident from oil leakage where gas escapes.
	Incorrect ignition timing	Check accuracy and reset if necessary.
Lack of power	Fault in fuel system or incorrect ignition timing	Check fuel lines or float chamber for sediment. Reset ignition timing.
Heavy oil consumption	Cylinder block in need of rebore	Check bore wear, rebore and fit oversize pistons if required.
Excessive mechanical noise	Damaged oil seals Worn cylinder bores (piston slap) Worn camshaft drive chain (rattle) Worn big-end bearings (knock) Worn main bearings (rumble)	Check engine for oil leaks. Rebore and fit oversize pistons. Adjust tensioner or replace chain. Fit replacement crankshaft assembly. Fit replacement crankshaft assembly.
Engine overheats and fades	Lubrication failure	Stop engine and check whether internal parts are receiving oil. Check oil level in crankcase.

Fault diagnosis for clutch and gearbox, overleaf

60 Fault diagnosis: clutch

Symptom	Cause	Remedy
Engine speed increases as shown by tachometer but machine does not respond	Clutch slip	Check clutch adjustment for free play, at handlebar lever, check thickness of inserted plates.
Difficulty in engaging gears, gear changes jerky and machine creeps forward when clutch is withdrawn, difficulty in selecting neutral	Clutch drag	Check clutch for too much free-play. Check plates for burrs on tongues or drum for indentations. Dress with file if damage not too great.
Clutch operation stiff	Damaged, trapped or frayed control cable	Check cable and replace if necessary. Make sure cable is lubricated and has no sharp bends.

61 Fault diagnosis: gearbox

Symptom	Cause	Remedy
Difficulty in engaging gears	Selector forks bent Gear clusters not assembled correctly	Replace with new forks. Check gear cluster for arrrangement and position of thrust washers.
Machine jumps out of gear	Worn dogs on the ends of gear pinions	Renew worn pinions.
Gear change lever does not return to original position	Broken return spring	Renew spring.
Kickstart does not return when engine is turned over or started (GN model only)	Broken or wrongly tensioned return spring	Renew spring or retension.
Kickstart slips (GN model only)	Ratchet assembly worn	Dismantle engine and replace all worn parts.

Chapter 2 Fuel system and lubrication

Refer to Chapter 7 for information relating to the 1981 to 1988 models

Contents

Specifications

Fuel tank capacity

Overall	22 lit (5.8/4.8 US/Imp gal)
Reserve	4.2 lit (1.1/0.9 US/Imp gal)
GS 850 GLT:	
Overall (No reserve capacity)	13 lit (3.4/2.9 US/Imp gal)

Fuel grade ....

Unleaded or low-lead, minimum octane rating 90 RON/RM

Carburettors

	GS 850 GN (UK)	GS 850 GN (US)	GS 850 GT/GLT (US)	GS 850 GT (UK)
Make	Mikuni	Mikuni	Mikuni	Mikuni
Type	VM26SS	VM26SS	BS32SS	BS32SS
ID No	45060	45100	45110	45120
Venturi size	26 mm	26 mm	32 mm	32 mm
Main jet	102.5	102.5	115	115
Main air jet	N/A	N/A	1.7	1.7
Jet needle – clip position, grooves from top	5DL36-2nd	5DL36-2nd	5D50	5D57-3rd
Needle jet	0–6	0–4	X-5	X-6
Pilot jet	15	15	40	42.5
Pilot air jet	1.2	1.2	190	200
Throttle valve	1.5	1.5	130	130
Starter jet	N/A	N/A	32.5	32.5
Float height	23.0–25.0 mm (0.91–0.98 in)	23.0–25.0 mm (0.91–0.98 in)	22.4±1.0 mm (0.88±0.04 in)	22.4±1.0 mm (0.88±0.04 in)
Pilot air screw	$1\frac{1}{4}$ turns out	Pre-set*	N/A	N/A
Pilot mixture screw	$\frac{5}{8}$ turn out	Pre-set*	Pre-set*	$3\frac{1}{2}$ turns
Idle speed	950–1150 rpm	950–1150 rpm	950–1150 rpm	950–1150 rpm
Fuel level	3–5 mm (0.12–0.20 in)	3–5 mm (0.12–0.20 in)	5±0.5 mm (0.20±0.02 in)	5±0.5 mm (0.20±0.02 in)

*Do not adjust

Lubrication system

Engine/transmission capacity:	
Dry	3.8 lit (8.0/6.6 US/Imp pint)
Without filter change	2.8 lit (5.9/4.9 US/Imp pint)
With filter change	3.6 lit (7.6/7.4 US/Imp pint)
Secondary gear capacity	340–400 cc (11.5–13.5/12.0–14.1 US/Imp oz)
Final drive gear capacity	280–330 cc (9.5–11.5/9.9–11.6 US/Imp oz)
Oil specification:	
Engine/transmission oil	SAE 10W/40 engine oil, API class SE or SF
Secondary and final drive gear oil	SAE 90 hypoid gear oil, API class GL-5 (SAE 80 gear oil if ambient temperature is below 0°C (32°F)

Oil pump:
    Type ....   Trochoid
    Output pressure ....   Above 0.1 kg/cm^2 (1.42 psi) and below 0.5 kg/cm^2 (7.1 psi) @
                                                   3000 rpm
    Outer rotor/housing clearance (max) .... 0.25 mm (0.010 in)
    Inner rotor/outer rotor clearance (max) .... 0.20 mm (0.008 in)
    Side clearance (max) .... 0.15 mm (0.006 in)

1 General description

The fuel system comprises a petrol tank, from which petrol is fed by gravity to the float chamber of each of the four carburettors, via a petrol tap. The single, vacuum type, petrol tap which incorporates a detachable gauze filter is fitted on the left-hand side of the tank. A tap of similar design is fitted to both the GS 850 GN and GT models, containing provision for a reserve quantity of petrol when the main supply is exhausted. The GS 850 GLT model, however, incorporates an unusual design of petrol tap in that there is no provision for a reserve supply of petrol and indeed, there is no actual petrol tap lever.

The petrol tap fitted to the GN and GT models has three positions; ON, RES, and PRI. Fuel can, however, only flow to the carburettors, when the engine is running, with the tap in the ON or RES position. This is due to the tap diaphragm which is controlled by induction pressure, and gives rise to the term 'vacuum operated' usually applied to this type of petrol tap. If there is no fuel in the float chambers, as may be the case after carburettor dismantling, the petrol tap should be turned to the priming position, to allow an unrestricted flow of petrol to the float chambers. Return the tap to the ON position, or if the fuel level is low, to the RES position, as soon as the engine is running. On the GLT model, the petrol tap is again vacuum controlled, but is of a simpler design. It has a two-position setting; ON or PRI. Unlike the more usual taps, the PRI setting is obtained by using a screwdriver in the adjuster slot provided; there is no lever on the tap and no reserve tank facility.

On the GS 850 GN models, four Mikuni throttle slide carburettors of VM26SS type are fitted. On the GT/GLT models, these have been replaced by four Mikuni BS32SS instruments of the constant depression (CD) type. On all models, the carburettors are mounted as a unit on a cast aluminium alloy bracket. The throttle slide type carburettors are controlled by a push-pull cable arrangement secured to a cross-rod, which passes through the top of each carburettor, connecting each throttle valve by a bell-crank arrangement. A similar operating mechanism is employed on the CD carburettors but controlled by a single cable.

For cold starting, a remotely controlled cable-operated choke mechanism is fitted, the control knob of which is mounted below the instrument panel. This enables the mixture entering the engine to be temporarily richened. When the engine has started, the choke can be gradually dispensed with as the engine warms up, until full air is accepted under normal running conditions, when the control knob can be pushed fully home.

Lubrication is effected by the wet-sump principle in which oil is delivered under pressure from the reservoir of oil contained within the engine sump, through a mechanical pump to the working parts of the engine. This oil is shared by the engine, primary drive, clutch and gearbox components. The oil pump is of the Eaton trochoid type and is driven from a pinion engaged with and to the rear of the clutch. Oil is supplied under pressure via a full flow oil filter fitted with a replaceable paper element to the crankshaft and to the overhead camshaft and valve gear. A secondary flow passes to the gearbox via the gearbox main bearings. All surplus oil drains to the sump by gravity where the distribution sequence is repeated. The oil pump itself is protected by a gauze strainer in the base of the oil pick-up chamber.

The secondary drive casing which contains the 90° bevel drive and driven gear units, holds its own reservoir of oil, which is remote from that of the engine. The somewhat different requirements of the bevel gear units, dictates the use of a high pressure hypoid-type gear oil to ensure correct and lasting lubrication.

2 Petrol tank: removal and replacement

1   The petrol tank is retained by two guide channels which locate with a circular rubber block on each side of the steering head and a bolt passing through a lug at the rear of the tank and into the frame.
2   To remove the tank, leave the petrol tap lever at the ON or (where fitted) the RESERVE position and detach the diaphragm vacuum pipe and the larger bore petrol feed pipe. The latter pipe is secured at the union by a spring clip, the ears of which should be pinched together to release the tension on the pipe. Remove the bolt from the rear of the tank, after raising or removing completely (GS 850 GLT model) the dual seat to gain access.
3   Detach the leads for the fuel gauge at the snap connectors below the left-hand side of the petrol tank.
4   Lift the tank up at the rear and then ease the unit backwards until the location cups leave the rubber blocks. At this stage the tank is obstructed by the location cups coming in contact with the ignition coils bolted to the frame. Lift the tank upwards off the machine.
5   Drainage of petrol is not strictly necessary when removing the tank, although the reduction in weight will facilitate the operation. A full tank will weigh approximately 35 lbs.
6   Replace the petrol tank by reversing the removal procedure. Take care not to trap control cables or stray wires between the tank and frame tubes. If the cups are a tight fit on the rubber blocks, apply a small amount of washing-up detergent to the blocks to ease refitting.

3 Petrol tap: removal and replacement

1   Removal of the complete petrol tap is required periodically to gain access to the filter columns for cleaning. The tank should be drained of petrol by fitting a length of tubing to the petrol tap outlet and turning the tap lever to the priming position. On the GLT model, insert a screwdriver blade into the adjuster slot on the petrol tap to obtain the priming position. Alternatively, if there is only a small quantity of fuel in the tank, the tank can be rested on its right-hand side so that all the petrol is away from the tap and will not spill out as the tap is removed.
2   The petrol tap is held to the underside of the petrol tank by two crosshead screws with washers. Note that there is an O-ring seal between the petrol tap body and the petrol tank, which must be renewed if it is damaged or petrol leakage has occurred. The filter screens which are integral with the plastic level pipes should be cleaned of any deposits using a soft brush and clean petrol. Because there is only a single tap to feed four carburettors, any restriction in petrol flow may lead to fuel starvation, causing missing, and in extreme cases overheating due to a weak mixture.
3   It is seldom necessary to remove the lever which operates the petrol tap on those models fitted with such a device. Should leakage develop at the joint, it will be necessary to drain the tank, or use the alternative procedure noted in the preceding paragraph, before the lever assembly can be removed. There is, however, no need to disturb the main body of the tap.
4   To dismantle the lever assembly, remove the two crosshead screws passing through the plate on which the operating positions are inscribed. The plate can then be lifted away,

followed by a wave washer, the lever itself and then the sealing gasket behind the lever. The sealing gasket will have to be renewed if leakage has occurred. Reassemble the tap in the reverse order. Gasket cement or any other sealing medium is not necessary to secure a petrol light seal.

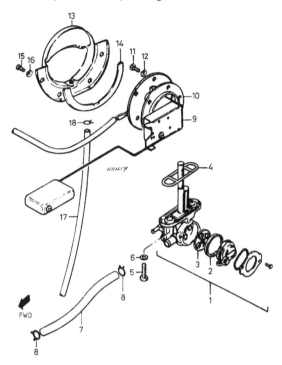

Fig. 2.1 Fuel tap and gauge – GN and GT models
Refer to Fig. 7.3 for GLT fuel tap

1 Fuel tap assembly	10 Sealing ring
2 Tap lever gasket	11 Bolt – 5 off
3 Tap valve gasket	12 Spring washer – 5 off
4 O-ring	13 Fuel drain plate
5 Bolt – 2 off	14 Gasket
6 Sealing washer – 2 off	15 Screw – 4 off
7 Fuel delivery pipe	16 Spring washer – 4 off
8 Clip – 2 off	17 Overflow pipe
9 Fuel level gauge	18 Clip

4 Carburettors: removal from the machine

1 To improve access to the carburettors it is suggested that the petrol tank is removed, as described in Section 2 of this Chapter, before dismantling proper commences.
2 Before detaching the air filter box or carburettors, displace the various vent and drain tubes from the securing clips.
3 Detach the engine breather hose from the unions at the air filter box and the breather cover on the cylinder head. The hose is secured at both ends by spring clips. On the GS 850 GN model, disconnect the two throttle cables from the operating pulley at the carburettors. Both may be detached in a similar manner. Loosen the upper and lower locknuts on the cable adjuster screw and displace the adjuster and outer cable from the abutment bracket. Rotate the pulley until the inner cable nipple can be pushed out of the anchor point. On the GT/GLT models, disconnect the single throttle cable from the operating pulley on the throttle control rod by pulling the outer cable up and forwards from the cable abutment, and then detaching the nipple from the pulley. Similarly, disconnect the choke operating cable from its holder and adjustment bracket.
3 Loosen the screw clips which secure the air filter hoses and inlet stubs to the carburettors. Remove the two mounting bolts

from the top of the air filter box and ease the box rearwards, so that the hoses leave the carburettor mouths.
4 Remove the air filter box from the machine to the right-hand side, and then detach the carburettors as a complete unit.

2.1 A single bolt retains the rear of the petrol tank

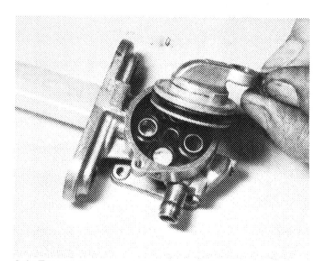
3.4a Two small screws retain tap lever to the tap body

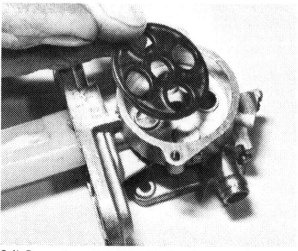
3.4b Renew the sealing gasket if leakage has occured

5 Carburettors: dismantling and reassembly – GS 850 GN model

1   The carburettors are mounted on a cast aluminium bracket which also serves as a support for the choke operating link rod and the cable anchor bracket. The bracket is so arranged that partial dismantling of all the carburettors is required before they can be removed from the bracket and attended to as individual items. Whenever possible, dismantle the carburettors separately, to prevent the accidental transposition of parts.

2   Remove the tops from the four carburettors. Each top is retained by three screws. Remove the bolt which passes through the forward end of each bellcrank and locates with the throttle link shaft. The throttle shaft is located longitudinally by a claw plate, secured by a single screw to a lug projecting from the mounting bracket. Remove the claw plate and then prise out the outer blind grommets from the left-hand and right-hand carburettors. Using a pair of snipe nosed pliers, disconnect the throttle pulley return spring from the two anchor pegs. The shaft can now be pushed out of position to free the bell cranks and the pulley wheel. Note that the pulley cannot be removed completely until the carburettors are separated.

3   Place the carburettors rear face downwards and remove the two countersunk screws which hold each instrument to the mounting bracket. Lift the bracket away, if necessary moving the choke link rod slightly so that the operating arm forks clear the choke plungers.

4   The carburettors are now joined only by the fuel cross feed pipes, which are a push fit in the bodies. Before separating the individual units, mark each carburettor carefully so that on reassembly no confusion arises as to their correct positions.

5   Select one carburettor and continue dismantling as follows, following suit with the other three carburettors, in turn. Lift out the bell crank, link arm and throttle valve unit, taking care not to damage the throttle needle. The throttle valve and needle may be removed from the valve seat by removing the two tiny screws. Invert the valve and allow the needle and needle clip to fall out. Removal of the link arm from the bell crank requires that the throttle slide vertical position adjuster screw is unscrewed fully and hence the original adjustment will be lost. It is unlikely that the link arm and bell crank will require separation and it is therefore advised that these components are left as a sub-assembly.

6   Invert the carburettor and remove the four screws that hold the float chamber to its base. Remove the hinge pin that locates the twin float assembly in each carburettor, and lift away the float. This will expose the float needle. The needle is very small and should be put in a safe place so that it is not misplaced.

7   Make sure the float chamber gasket is in good condition. It should not be disturbed unless it shows signs of damage or has been leaking.

8   Unscrew the main jet from the jet holder, using a wide bladed screwdriver and then unscrew the holder itself. Note the O-ring fitted above the holder threads. Invert the carburettor body and displace the needle jet from the central bore.

9   Unscrew the pilot jet from the boss to the side of the main jet housing.

10   Note that on the US market GN model the pilot air adjusting screw **must not** be removed or adjusted. On these models, the screw is pre-set under factory conditions and cannot be re-set, when once maladjusted, except under the same conditions. See Section 6 of this Chapter for information on carburettor adjustments in relation to US Emission regulations. Reference should be made to that Section before attempting any operation controlling the jetting and/or the mixture of the carburettors. The foregoing applies equally to the pilot mixture screw on all models. Do not touch them!

11   On the UK market GS 850 GN models, provision is made for adjusting the pilot air screw. Refer to Section 10 of this Chapter for the adjustment procedure, after reassembly.

12   The starter plunger (choke) assembly is positioned in a tunnel to the side of the upper chamber. Unscrew the housing cap and pull the starter plunger assembly out. This consists of the plunger rod, spring and plunger piece.

13   Check the condition of the floats. If they are damaged in any way, they should be renewed. The float needle and needle seating will wear after lengthy service and should be inspected carefully. Wear usually takes the form of a ridge or groove, which will cause the float needle to seat imperfectly. Always renew the seating and needle as a pair. An imperfection in one component will soon produce similar wear in the other.

14   After considerable service the throttle needle and the needle jet in which it slides will wear, resulting in an increase in petrol consumption. Wear is caused by the passage of petrol and the two components rubbing together. It is advisable to renew the jet periodically in conjunction with the throttle needle.

15   Before the carburettors are reassembled, using the reversed dismantling procedure, each should be cleaned out thoroughly using compressed air. Avoid using a piece of rag since there is always risk of particles of lint obstructing the internal passage-ways or the jet orifices.

16   Never use a piece of wire or any pointed metal object to clear a blocked jet. It is only too easy to enlarge the jet under these circumstances and increase the rate of petrol consumption. If compressed air is not available, a blast of air from a tyre pump will usually suffice.

17   Do not use excessive force when reassembling a carburettor because it is easy to shear a jet or some of the smaller screws. Furthermore, the carburettors are cast in a zinc-based alloy which itself does not have a high tensile strength. Take particular care when replacing the throttle valves to ensure the needles align with the jet seats.

18   Reassemble the carburettors by reversing the dismantling procedure. Before inserting the throttle link shaft, it should be lubricated with grease. Check the condition of the two O-rings which seal each side of the fuel transfer pipes. Renew them if there is any doubt as to their efficiency. Fit and tighten the eight carburettor mounting screws before tightening fully the through bolts which secure the bell cranks to the throttle shaft. A small amount of locking fluid should be applied to the mounting screw threads before they are inserted.

19   Before replacing the carburettors on the machine and before refitting the carburettor tops, refer to Section 8 for details of carburettor synchronisation.

6 Carburettors: Emission Control information – US models

1   All the GS 850 models marketed in the USA and introduced after January 1st 1978, are equipped with specially manufactured carburettors, containing certain components machined to extremely close tolerances. There are three specific components that are of a particularly close tolerance; the main jet, the needle jet and the pilot air jet. If replacement of any of these jets becomes necessary, a new part of the same, close tolerance type, must be obtained and fitted. To aid the operator in selecting the correct jet, the three later type jets use a different style of numerical identification. See accompanying table. The reason for the adoption of pre-set mixture adjustments and extremely close tolerance carburettor jetting, is ostensibly to enable the GS 850 to meet the US Federal Emissions Regulations. These regulations state that all the motor cycles produced after January 1st 1978, must comply with the strict statutory level of emissions. Therefore, by adjusting, resetting, or replacing the pilot air or pilot mixture screw, the emission levels may be adversely affected, thus leading to a possible infringement of State or Federal emission regulations. This could subsequently render the operator subject to a large fine. It is, therefore, strongly recommended that the advice of your local Suzuki dealer be sought, before attempting any operation controlling the jetting and/or the mixture of the carburettors.

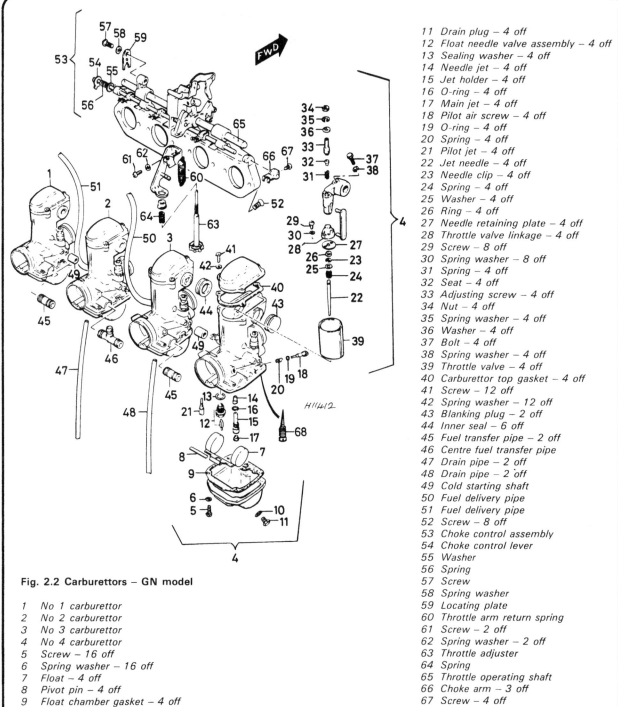

Fig. 2.2 Carburettors – GN model

1   No 1 carburettor
2   No 2 carburettor
3   No 3 carburettor
4   No 4 carburettor
5   Screw – 16 off
6   Spring washer – 16 off
7   Float – 4 off
8   Pivot pin – 4 off
9   Float chamber gasket – 4 off
10   Sealing washer – 4 off

11   Drain plug – 4 off
12   Float needle valve assembly – 4 off
13   Sealing washer – 4 off
14   Needle jet – 4 off
15   Jet holder – 4 off
16   O-ring – 4 off
17   Main jet – 4 off
18   Pilot air screw – 4 off
19   O-ring – 4 off
20   Spring – 4 off
21   Pilot jet – 4 off
22   Jet needle – 4 off
23   Needle clip – 4 off
24   Spring – 4 off
25   Washer – 4 off
26   Ring – 4 off
27   Needle retaining plate – 4 off
28   Throttle valve linkage – 4 off
29   Screw – 8 off
30   Spring washer – 8 off
31   Spring – 4 off
32   Seat – 4 off
33   Adjusting screw – 4 off
34   Nut – 4 off
35   Spring washer – 4 off
36   Washer – 4 off
37   Bolt – 4 off
38   Spring washer – 4 off
39   Throttle valve – 4 off
40   Carburettor top gasket – 4 off
41   Screw – 12 off
42   Spring washer – 12 off
43   Blanking plug – 2 off
44   Inner seal – 6 off
45   Fuel transfer pipe – 2 off
46   Centre fuel transfer pipe
47   Drain pipe – 2 off
48   Drain pipe – 2 off
49   Cold starting shaft
50   Fuel delivery pipe
51   Fuel delivery pipe
52   Screw – 8 off
53   Choke control assembly
54   Choke control lever
55   Washer
56   Spring
57   Screw
58   Spring washer
59   Locating plate
60   Throttle arm return spring
61   Screw – 2 off
62   Spring washer – 2 off
63   Throttle adjuster
64   Spring
65   Throttle operating shaft
66   Choke arm – 3 off
67   Screw – 4 off
68   Pilot mixture screw – 4 off

Conventional Figures Used On Standard Tolerance Jet Components	1 2 3 4 5 6 7 8 9 0
Emission Type Figures Used On Close tolerance Jet Components	1 2 3 4 5 6 7 8 9 0

Fig. 2.3 Numerals used on carburettor jets

5.2a Remove bolt from each bellcrank and ...

5.2b ... detach the claw plate which secures the throttle shaft

5.2c Disconnect the throttle pulley return spring and ...

5.2d ... withdraw the throttle shaft

5.3a Remove all the countersunk screws and ...

5.3b ... lift the mounting bracket away to separate the carburettors

5.4 Separate the fuel cross feed pipes; carburettors are now free

5.5a Withdraw the throttle valve unit

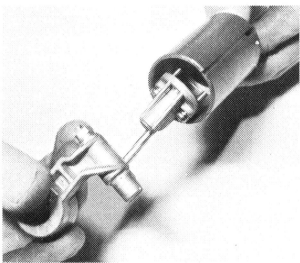

5.5b Remove crank link rod from throttle valve having unscrewed the ...

5.5c ... two small securing screws

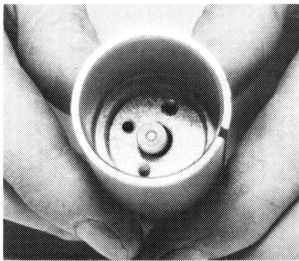

5.5d Note the nylon sealing ring on the throttle needle upper end

5.6a Float chamber is held by four screws (two shown). Note pilot mixture screw location (arrowed)

5.6b Removal of the screws allows float chamber to be separated from its base

5.6c Push out pivot pin to detach float assembly. Support the pin columns whilst doing this

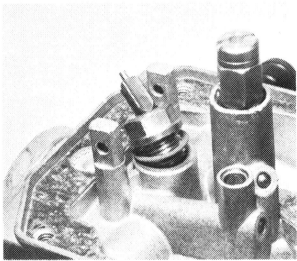

5.6d Do not lose the minute float needle. The seat may be unscrewed for renewal

5.7 Check that the float chamber gasket is in good condition

5.8a Main jet may be unscrewed from its holder

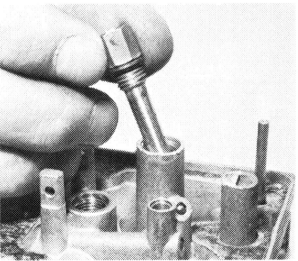

5.8b Note the O-ring on the main jet holder

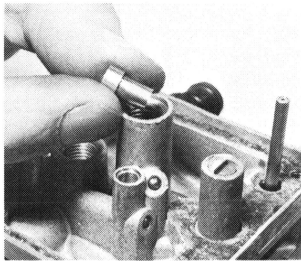

5.8c Remove the needle jet from the central bore

5.9 Unscrew pilot jet from boss to side of main jet housing

5.10 **Do not** adjust pilot air adjusting screw (see text)

5.12a If, as may rarely occur, choke plunger gives trouble, separate plunger from lever arm ...

5.12b ... and withdraw plunger assembly for examination

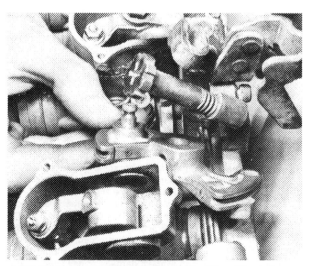

5.18a When reassembling, ensure the throttle cables pulley wheel assembly is positioned before ...

5.18b ... reinserting the throttle link shaft

5.18c Secure choke link arm with their screws

7 Carburettors: dismantling and reassembling – GS 850 GT/GLT models

1   The carburettors are mounted on two angled cast aluminium brackets; one runs along the lower, forward side of the carburettors unit, the other along the rear, upper side of the unit. In addition, two smaller brackets are fitted to the two central carburettors. One of these brackets acts as the throttle cable holder and the other holds the choke cable in position. The bracket which holds the throttle cable in position is retained to the No 2 carburettor by two of the crosshead screws that secure the diaphragm cover to the carburettor top. The choke cable holder bracket is secured by one screw which also serves as one of the four diaphragm cover retaining screws on carburettor No 3.
2   Remove the three screws which retain the throttle cable and choke cable holding brackets, and detach the brackets. Loosen the four screws in the choke operating arm levers to enable the choke operating link rod to be removed. Pull the link rod clear of the carburettors. As the link rod is freed, the choke operating arm levers, through which the link rod passes, can all be detached.
3   Remove all the retaining screws from the two angled mounting brackets, and detach the brackets from the carburettors. The four carburettors can now be separated; they are now only joined by the fuel cross-feed pipes, which are a push fit. Before dismantling the carburettors ensure that they are carefully marked so that on reassembly no confusion arises as to their correct positions.
4   Remove the nut that retains the idle adjusting screw (throttle stop screw) and detach the screw. Remove three screws and detach the bracket on which the throttle stop screw is mounted.
5   Select one carburettor and continue dismantling as follows, following suit with the other three carburettors in turn.
6   Invert the carburettor and remove the four crosshead screws that retain the float chamber to its base. Remove the pivot pin that locates the twin float assembly in each carburettor, and lift away the float assembly. The pivot pin is an interference fit at the headed end in the support post and must, therefore be drifted from place. Ensure that the post is supported when drifting out the pin or the post will snap off. This will expose the float needle. The needle is very small and should be put in a safe place so that it is not misplaced. Make sure that the float chamber gasket is in good condition. It should not be disturbed unless it shows signs of damage or has been leaking.
7   Check that the twin float assembly is in good condition and

7.5 Use the throttle stop screw for tick-over speed adjustments, with the carburettors installed on the engine

not punctured. Due to its construction, it is impossible to effect a repair should puncturing have occurred. The float must be renewed to restore correct operating efficiency.
8   Unscrew the main jet from the mixing chamber, using a close fitting, wide-bladed screwdriver. The needle jet is a push fit below the main jet, being retained by a small washer and the main jet. Check the needle jet for wear, together with the jet needle. After lengthy service, these two components should be renewed together, or high petrol consumption will result.
9   The float needle will also wear after lengthy service, and should be closely examined with a magnifying glass. Wear takes the form of a ridge or groove, which will cause the float needle to seat imperfectly. The needle and seating should always be renewed as a pair. The seating is a screw fit in the mixing chamber. Note the O-ring and also the tiny filter gauze, which is retained by the seat.
10   Remove the screws from the top of the carburettor. Lift the top from position, together with the piston spring. Carefully lift the diaphragm from position, bringing with it the piston and jet needle. Carefully check the condition of the diaphragm. If it has developed cracks or holes, it must be renewed as a unit, with the piston. The jet needle is retained by a guide, holder and clip. The jet needle must be renewed if worn, as described in paragraph 8. Note that the plastic retainer above the jet needle has a small protrusion which, when installed, skews the needle. The needle should be skewed towards the engine.

11 If any of the main mixture control components (the main jet, the needle jet and the pilot air jet) need replacing, reference should be made to Section 6, before any replacement parts are obtained. Do not remove or adjust the pilot mixture screw on US market models.

12 Before reassembling the carburettors, by reversing the dismantling procedure, refer back again to paragraphs 15 and 16 of the previous Section for carburettor cleaning procedure.

13 Do not use excessive force when reassembling a carburettor because it is easy to shear a jet or some of the smaller screws. Furthermore, the carburettors are cast in a zinc-based alloy, which itself does not have a high tensile strength.

14 When fitting the piston unit note that the diaphragm periphery has a location tab which locates in a slight recess in the main body.

15 Do not omit the fuel transfer T-pieces when replacing the carburettors on the two mounting brackets. If the pipes are not a good push fit, they may leak and should therefore be renewed.

The link arm lever on each throttle valve, through which passes the throttle operating shaft, should be fitted so that the lever is between the spring-loaded plunger and the throttle adjusting screw on the link arm of the next carburettor.

16 The eight crosshead screws used on each of the two mounting brackets should all be coated with a small amount of locking fluid prior to reassembly. When the choke operating link rod has been inserted and the four link rod levers have been replaced, secure each lever with a screw treated with a coat of locking fluid.

17 After assembly of the carburettors on the mounting brackets, check that all four throttle valves open fully and simultaneously. If differences occur, adjustment should be made using the throttle adjustment screws on the throttle shaft link arms. The separate idle adjuster screw with the notched head, should be used only to set the throttle opening when adjustng the tickover (idle) speed with the engine running.

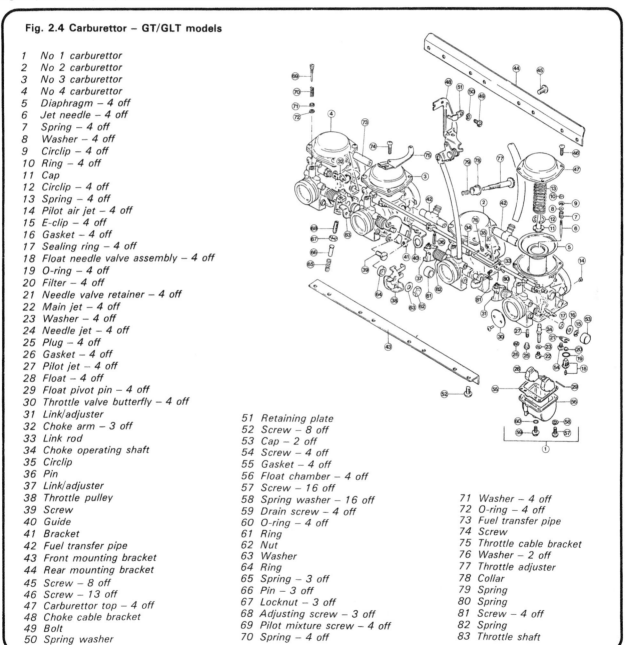

Fig. 2.4 Carburettor – GT/GLT models

1   No 1 carburettor
2   No 2 carburettor
3   No 3 carburettor
4   No 4 carburettor
5   Diaphragm – 4 off
6   Jet needle – 4 off
7   Spring – 4 off
8   Washer – 4 off
9   Circlip – 4 off
10 Ring – 4 off
11 Cap
12 Circlip – 4 off
13 Spring – 4 off
14 Pilot air jet – 4 off
15 E-clip – 4 off
16 Gasket – 4 off
17 Sealing ring – 4 off
18 Float needle valve assembly – 4 off
19 O-ring – 4 off
20 Filter – 4 off
21 Needle valve retainer – 4 off
22 Main jet – 4 off
23 Washer – 4 off
24 Needle jet – 4 off
25 Plug – 4 off
26 Gasket – 4 off
27 Pilot jet – 4 off
28 Float – 4 off
29 Float pivot pin – 4 off
30 Throttle valve butterfly – 4 off
31 Link/adjuster
32 Choke arm – 3 off
33 Link rod
34 Choke operating shaft
35 Circlip
36 Pin
37 Link/adjuster
38 Throttle pulley
39 Screw
40 Guide
41 Bracket
42 Fuel transfer pipe
43 Front mounting bracket
44 Rear mounting bracket
45 Screw – 8 off
46 Screw – 13 off
47 Carburettor top – 4 off
48 Choke cable bracket
49 Bolt
50 Spring washer

51 Retaining plate
52 Screw – 8 off
53 Cap – 2 off
54 Screw – 4 off
55 Gasket – 4 off
56 Float chamber – 4 off
57 Screw – 16 off
58 Spring washer – 16 off
59 Drain screw – 4 off
60 O-ring – 4 off
61 Ring
62 Nut
63 Washer
64 Ring
65 Spring – 3 off
66 Pin – 3 off
67 Locknut – 3 off
68 Adjusting screw – 3 off
69 Pilot mixture screw – 4 off
70 Spring – 4 off

71 Washer – 4 off
72 O-ring – 4 off
73 Fuel transfer pipe
74 Screw
75 Throttle cable bracket
76 Washer – 2 off
77 Throttle adjuster
78 Collar
79 Spring
80 Spring
81 Screw – 4 off
82 Spring
83 Throttle shaft

8   Carburettors: pilot screw adjustment

1   If the engine starts poorly and does not respond correctly to the use of the choke control, idles roughly and erratically, and if the transition from idle to higher engine speeds is not completely smooth, then it is possible that the pilot mixture settings require adjustment. *The pilot screw(s) should not be disturbed unless absolutely necessary and the final settings should never deviate far from standard unless aftermarket accessory equipment has been fitted which affects the engine's performance.*

2   Before disturbing the carburettors **always** check all other systems first. The engine must be in good condition, ie the air filter element must be clean, with no leaks in the filter assembly or hoses, the carburettors must be clean, securely fastened and have correct fuel levels, and the ignition timing and spark plugs correctly adjusted. Check the valve clearances and engine compression pressures to ensure that the engine is mechanically sound and ensure that the exhaust system is in good condition and is securely fastened, with no leaks. Start the engine and warm it up to normal operating temperature then set the idle speed to the specified amount.

3   On GN models the pilot mixture setting is controlled by an air screw, fitted at 45° downwards into the air filter end of each carburettor, and also by a mixture screw which projects vertically upwards at the front edge of each float chamber. On GT/GLT models there is only a mixture screw, projecting vertically downwards into each carburettor's body, immediately in front of the diaphragm cover. On later US models the screws may be sealed by a small cap or plug to prevent tampering.

4   To achieve the standard setting, gentle turn each screw inwards (clockwise) until it seats lightly, then unscrew it by the specified number of turns. Where a setting is given as 'Preset', the screw **must not** be disturbed except for cleaning (if required); if this is the case count the number of turns necessary to seat the screw lightly from the existing setting, unscrew it and reverse the procedure on reassembly to return the screw to its previous position.

5   **Do not** alter any pilot screw setting from standard. The normal procedure is to turn each screw, by a small amount at a time, each way from the standard setting until the position is found where the engine runs fastest and smoothest when fully warmed up, although in practice it is not possible to distinguish significant differences without an exhaust gas analyser and a very sensitive tachometer. Owners of US models should also note the possible legal consequences of attempting to alter the mixture setting; see Section 6. If the pilot mixture is thought to require adjustment, the machine should be taken to a good Suzuki Service Agent for the work to be done by an expert using equipment of the required scope and accuracy.

9   Carburettors: synchronisation – GS 850 GN models

1   On UK market models the pilot air screw adjustment procedure should be carried out as described in the preceding Section prior to synchronising the carburettors. US market models have pre-set pilot air screws and no attempts at adjustment should be made. Synchronisation of the carburettors should be carried out in two stages. The first stage as detailed in the following paragraph should be accomplished with the carburettors removed from the machine, at any time after the carburettors have been dismantled and reassembled. The second stage, which is synchronisation of the carburettors using either a balancer device or dial-type vacuum gauge, should be carried out as a routine maintenance item. It is also necessary when rough idling or poor engine performance is encountered, or after the carburettors have been refitted to the machine.

2   Unscrew the remote throttle stop screw, which is fitted with a large notched head, so that clearance can be seen between the end of the screw and the portion of the throttle pulley against which it normally abuts. The throttle slides should now be in the fully closed position. Make a visual check that all four throttle slides open and close at precisely the same time. If variations occur, each slide may be adjusted individually by means of the adjuster screw on the end of the bell crank. Loosen the locknut on the screw and make the required adjustment. Retighten the locknut without moving the screw. Rotate the throttle pulley so that the slides are in the fully open position and check that the lower edges of the slides are at the position indicated in the accompanying diagram. Adjustment of this setting should be made by turning the single, spring secured screw, which is fitted to the rear face of the throttle cable anchor plate. After making the fully open and fully closed adjustment, replace the carburettor tops and refit the carburettor and controls to the machine.

3   Adjust the throttle cables by starting with the opening cable first. Loosen the locknut on the throttle opening cable, and use the adjuster to allow 1.0 – 1.5 mm (0.04 – 0.06 in) of slack in the cable before securing the locknut again. Loosen the locknut on the closing cable, and adjust it so that there is no play in the throttle twist-grip, then secure the locknut.

4   As stated above, running adjustment of the carburettors requires the use of specialised equipment. The adjustment must be made using either the correct Suzuki carburettor balancer (Part No. 09913–13121) or by using a set of dial-type vacuum gauges. When using either method the appropriate adaptors, which screw into the inlet tracts and to which are attached the vacuum pipes, must also be obtained. Unless the carburettor vacuum balancer device or the dial-type vacuum gauge set is to hand, it is recommended that the machine is returned to a Suzuki Service Agent, who will carry out the synchronisation as a normal service task.

5   The Suzuki vacuum balancer system operates on the principle of a steel ball rising and falling within a column. The four columns each contain one steel ball. The columns carry three horizontal lines each, which, when related to the position of the steel balls, serve to indicate the 'balance' of the carburettors.

6   Remove the hexagon headed blanking screws from the inlet tracts on the cylinder head with the use of a 4 mm Allen key. If the correct Suzuki balancer is to be used, initially only remove the blanking screw from either the No 1 or No 4 carburettor inlet tract. Insert one of the adaptors into the blanking screw hole which has been revealed and attach one of the outer vacuum tubes on the balancer to the adaptor. Start the engine and allow it to run at a steady speed of between 1500 – 2000 rpm. It will be noted that in the base of the balancer, below each column, is an adjuster screw. These screws act as an air screw and by screwing them inwards or outwards, the vacuum in the relevant column is raised or lowered. Using the adjuster screw of the column in question, adjust the vacuum in that column until the steel ball lies directly below the central horizontal line. With this accomplished, disconnect the vacuum tube from the adaptor and connect the next tube on the gauge to the same adaptor on the same carburettor. The steel ball in this next column should again be adjusted to the central position by means of the relevant adjuster screw. Repeat the procedure for the remaining two carburettors using the relevant vacuum tubes and columns. In this way a datum carburettor (either No. 1 or No. 4) is obtained. It is now possible to synchronise all four carburettors simultaneously by adjusting the three other carburettors to produce the same reading as that for the datum. Make adjustment as follows: remove all the blanking screws from the inlet tracts, insert the four adaptors and attach the four vacuum tubes from the balancer gauge to the adaptors. Start the engine and raise the engine speed to a steady speed between 1500 – 2000 rpm. If the four carburettors are correctly balanced, ie in proper synchronisation, the four steel balls will all be located in line with the central line in their respective columns.

7 If adjustment is now required, the tops of the carburettors requiring adjustment should be detached. Each top is retained by three screws. To facilitate detachment of the carburettor tops, release the rear of the petrol tank from its single mounting bolt and disconnect the petrol feed pipe and vacuum pipe at the petrol tap. Also disconnect the two fuel gauge leads at their snap connectors. The tank can now either be removed completely or the rear section propped up to provide sufficient clearance; see Section 10, paragraph 4. To facilitate adjustment further, a special Suzuki service tool is available (No. 09913–14520). This tool is correctly shaped to enter the carburettor body and adjust the bell-crank adjuster screw by which means adjustment of the 'balance' is effected. The factory tool is, however, not a necessity as a small ring spanner (to release the adjuster screw locknut) and a short bladed screwdriver will enable adjustment to be carried out successfully. With adjustment of the carburettors complete and the synchronisation correct, refit the carburettor tops and the petrol tank. The engine idling speed will also probably need readjusting at this stage. Using the large headed throttle stop screw, adjust the idling speed to between 950 – 1150 rpm with the engine at normal operating temperature.

8 If the dial-type vacuum gauge method of carburettor synchronisation is to be used in preference to the Suzuki balance gauge set-up, a similar procedure should be adopted.

9 Remove the blanking screws from the inlet tracts on the cylinder head, fit the adaptors and connect up the vacuum gauges. Start the engine and allow it to run until normal running temperature has been reached. By means of the throttle stop screw raise the engine speed to a steady speed between 1500 – 2000 rpm. Select as a datum the carburettor which shows the central reading. Make adjustment using the bell-crank adjuster screw in each carburettor so that the three readings on the remaining gauges are modified to correspond with that of the datum gauge. Using the throttle stop screw reduce the engine speed to the specified tick-over speed of 1050 rpm. Disconnect the vacuum gauges and refit the take-off blanking plugs, together with their sealing washers.

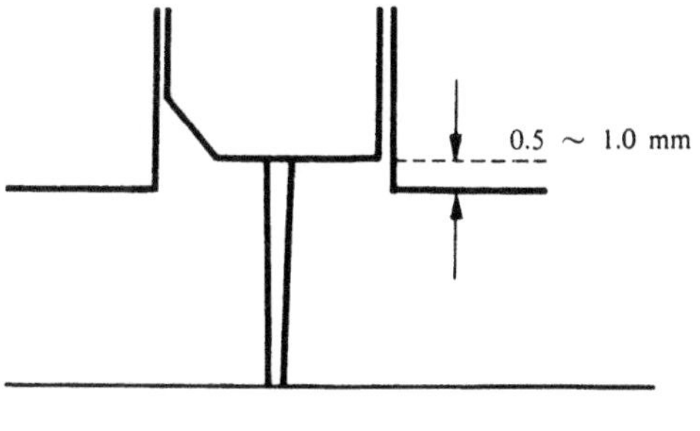

Fig. 2.5 Correct positioning of throttle valve

0.5 ~ 1.0 mm

10 Carburettors: synchronisation – GS 850 GT/GLT models

1 In general the details given in Section 9 for the GS850 models apply, note however that the procedure is complicated by the need to set the outer two carburettors at a higher level than the inner two to compensate for the different lengths of the inlet tracts between the inner and outer carburettors and the air filter element. Also either the official Suzuki tool or a set of good quality, glycerine-damped, dial-type gauges will be required. For this reason it is recommended that the machine be taken to a Suzuki Service Agent for the work to be carried out by an expert using the correct equipment. Those owners who wish to carry out the work themselves should proceed as follows.

2 If the carburettors have been dismantled and reassembled the throttle butterfly valves may be aligned visually so that they open and close at the same time as each other. The No 3 carburettor should be considered the datum carburettor as it is this instrument which is fitted with the main throttle stop screw. The other carburettors are fitted with adjuster screws and locknuts on the throttle spindle interconnecting links. This is a necessary preliminary to full vacuum synchronisation (balancing) and should be carried out whenever the engine is running poorly.

3 It must be noted that synchronisation cannot be carried out successfully unless all other adjustments are normal, especially the remaining carburettor settings. Unless these are set correctly, any attempt at synchronisation will produce erroneous readings. The engine must first be checked thoroughly to ensure that it is in good condition, ie that the air filter element is clean, that there are no leaks in the filter assembly or hoses, that the carburettor fuel levels and (where applicable) the pilot screw settings are correct, that the carburettors are clean and securely fastened and that the ignition timing and spark plugs are correctly adjusted. Check also the valve clearances and engine compression pressures to ensure that the engine is mechanically sound and ensure that the exhaust system is in good condition and is securely fastened, with no leaks. Start the engine and warm it up to normal operating temperature then set the idle speed to the specified amount.

4 Note that if the rear of the tank is raised or if it is removed completely to improve access to the balancing screws, either an alternative fuel supply must be arranged or the fuel and vacuum hoses must be temporarily extended. If the vacuum hose is disconnected it must be plugged securely to allow the engine to run and to ensure that the gauge readings are correct.

5 If using the Suzuki service tool follow the instructions given in Section 9, paragraph 6 to calibrate and use the equipment. Note that the steel balls of the gauges for the two inner cylinders (Nos 2 and 3) should be aligned with their top edges just touching the gauge centre lines, while those for the gauges of the two outer cylinders should be aligned so that they are bisected by the centre line, ie the two inner cylinders should show half of a ball diameter less vacuum than the outer two. If adjustment is necessary, proceed as described in paragraphs 10 and 11.

6 If using dial-type gauges first check that they are sufficiently accurate by connecting each gauge in turn to any one cylinder. If any gauge does not give exactly the same reading as its counterparts, or if any is inconsistent, then the complete set cannot be used. Where the damping is adjustable set it so that needle flutter is just eliminated, but so that the gauge can record the slightest change in pressure.

7 When the gauges are known to be accurate, connect each to its respective cylinder. Since Suzuki do not give the necessary information the difference in pressure between the inner and outer cylinder can only be approximated; the actual figure will vary greatly depending on the quality of the gauges used.

8 Since this difference in pressure is due to the presence of the air filter element, a reasonably accurate setting can be achieved by temporarily removing the element for the duration of the adjustment and then balancing all four cylinders to exactly the **same** level; if the element is then refitted the difference in pressure can be noted for future reference.

9 With the gauges connected and the element removed, therefore, check that all cylinders give the same reading and make the necessary adjustments as described in paragraphs 10 and 11. Do not forget to refit the filter element once adjustment is complete. **Note:** *this operation is described only as an alternative for those owners who have access to a set of dial-type gauges but not to the Suzuki service tool. The results obtained can only approximate the correct setting and will depend on the accuracy of the gauges used and the skill of the owner. If there is any doubt about the results, the machine must be taken to a Suzuki Service Agent for the carburettor synchronization to be checked using the correct tool.*

10 **Adjustment.** If adjustment is necessary a specific sequence must be followed; note that the cylinders are numbered

1, 2, 3 and 4 in sequence from left to right and that all are balanced against No. 3 cylinder. Use the centre screw first to match No. 2 cylinder with No. 3, then the left-hand screw to set No. 1 cylinder. Finally use the right-hand screw to set No. 4 cylinder. Do not press down on the screws when making adjustments and ensure that the setting does not alter when the locknuts are tightened. After each adjustment open and close the throttle quickly to settle the linkage, wait for the gauge reading to stabilise and note the effect of the adjustment before proceeding; the idle speed and vacuum reading may vary noticeably.

11 During the course of adjustment do not allow the engine to overheat; stop it and wait for it to cool down if necessary. When the carburettors are correctly synchronised stop the engine, disconnect the gauges and refit the fuel tank and blanking plugs. Check the idle speed and throttle cable free play, resetting each if necessary.

12 **Note:** *when using either gauge, it should be noted that if aftermarket accessory filters have been substituted for the standard filter assembly and plenum chamber, some experimentation may be necessary to achieve the correct setting. Since the difference in inlet tract lengths will no longer exist, it is possible that the best results will be obtained with all cylinders set to the same level.*

11 Carburettors: checking the float chamber fuel level – all models

1   If conditions of a continual weak mixture or flooding are encountered on one or more carburettors, or if difficulty is experienced in tuning the carburettors, the float levels should be checked and, if necessary, adjusted. Although the float chambers maybe removed with the carburettors in situ on the machine, it is advised that the carburettors be removed to facilitate inspection and adjustment.

2   The float level is correct when the distance between the uppermost edge of the floats (with the carburettor inverted) and the mixing chamber body flange is 23 – 25 mm (0.90 – 0.98 in) on the GS 850 GN model, or 22.4 ± 1.0 mm (0.88 ± 0.04 in) on the GT/GLT models. Note that the gasket must be removed from the mixing chamber body before the measurement is taken. The floats should be in the closed position when the measurement is taken. Adjustment is made by bending the float assembly tang (tongue), which engages with the float, in the direction required.

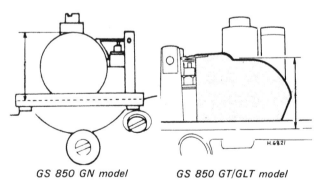

GS 850 GN model   GS 850 GT/GLT model

Fig. 2.6 Float height adjustment

12 Carburettors: settings

1   Some of the carburettor settings, such as the sizes of the needle jets, main jets, and needle positions are pre-determined by the manufacturer. Under normal riding conditions it is unlikely that these settings will require modification. If a change

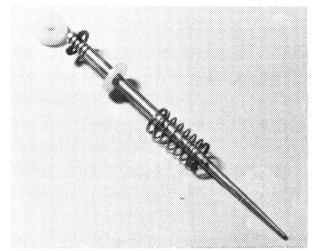

11.2 Mid-range mixture strength may be adjusted by varying the position of the jet needle

12.2 Remove screw to free element from filter box

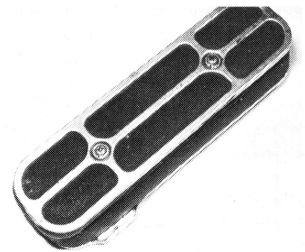

12.3 Two screws retain element in frame/carrier

appears necessary, it is often because of an engine fault, or an engine modification, eg a leaky exhaust port joint or an after-market exhaust system.

2   Apart from alterations of the pilot adjuster screws (GS 850 GN and GT models – UK variant), some adjustment of the mid-range mixture strength can be made on the GN model only (US and UK market), by raising or lowering the jet needle. This is accomplished by changing the position of the needle clip; fitting it into a different groove. Raising the needle will richen the mixture and lowering the needle will weaken the mixture.

3   Always err slightly towards a rich mixture as one that is too weak will cause the engine to overheat and burn the exhaust valves. Reference to Chapter 3 will show how the condition of the sparking plugs can be interpreted with some experience as a reliable guide to carburettor mixture strength.

13 Air cleaner: dismantling, servicing and reassembly

1   The air cleaner is mounted immediately behind the four carburettors into which the carburettor intakes fit. The air filter housing contains the element that is removable for cleaning or replacement, when necessary.

2   In order to gain access to the air filter element, there is no necessity to remove the air filter housing itself. The air filter box on all the GS 850 models is fitted with a chromed cover each side of the machine, both of which are retained by two screws and lock washers. Detach both chromed covers, and remove the screw(s) which retains the element frame/carrier. The element and its frame/carrier may be pulled or pushed out of the filter box lower section where it is located.

3   Detach the element from the carrier after removing the two retaining crosshead screws. The element is of the oil impregnated polyurethane sponge type. The element should be cleaned thoroughly in a non-flammable solvent (**do not** use petrol), squeezing it gently to dislodge any accumulated debris. Do not attempt to wring the foam element out as it is easily torn. The solvent should be removed by squeezing the foam carefully in some clean rag. Allow a short time to allow any residual solvent to evaporate.

4   Check the cleaned element carefully, looking for holes or splits, which if found, will necessitate the element's renewal. Immerse the cleaned element in engine oil, then squeeze it gently to remove any excess. The object is to leave it moist, but not dripping with oil.

5   Reinstall the element in its frame/carrier, and refit the assembly into the air filter box, by reversing the removal procedure. Check that the carrier locating tabs are positioned in their slots, so that the element is pressed firmly against the housing mating surface.

6   The air filter element should be removed for cleaning at approximately 6000 km (4000 miles) intervals. If the foam becomes damaged or hardened with age, it should be renewed as a matter of course.

7   Never run the machine without the element or with the air filter disconnected. The jetting of the carburettors takes into account the presence of the air filter, and the weak mixture that results from the absence of the filter will cause engine over-heating and eventually severe engine damage.

14 Engine and gearbox lubrication

1   As previously described at the beginning of the Chapter the lubrication system is of the wet sump type, with the oil being forceably pumped from the sump to positions at the gearbox bearings, the main engine bearings, and the cam box bearings, all oil eventually draining back to the sump. The system incorporates a gear driven oil pump, an oil filter, a safety by-pass valve, and an oil pressure switch. Oil vapours created in the crankcase are vented through a breather to the air cleaner box, where they are passed into the cylinder providing an oiltight system.

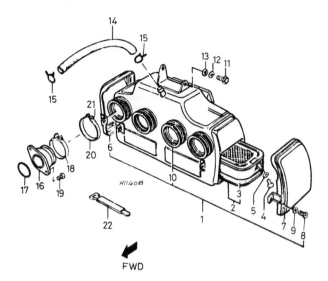

FWD

Fig. 2.7 Air cleaner assembly

1 Air cleaner assembly	12 Spring washer – 2 off
2 Air filter assembly	13 Washer – 2 off
3 Element	14 Breather pipe
4 Screw	15 Pipe clip – 2 off
5 Spring washer	16 Inlet hose – 4 off
6 Right-hand end cover	17 O-ring – 4 off
7 Left-hand end cover	18 Hose clamp – 4 off
8 Screw – 4 off	19 Screw – 8 off
9 Spring washer – 4 off	20 Hose clamp – 4 off
10 Air transfer hose – 4 off	21 Screw – 4 off
11 Bolt – 2 off	22 Cable clip

2   The oil pump is an Eaton trochoid twin rotor unit which is driven from a gear engaged with and to the rear of the clutch. An oil strainer is fitted to the intake side of the pump, which serves to protect the pump mechanism from impurities in the oil which might cause damage.

3   A corrugated paper oil filter is included in the system and is fitted within an enclosed chamber in the front of the crankcase. Access to the filter is made through a finned cover. As the oil filter unit becomes clogged with impurities, its ability to function correctly is reduced, and if it becomes so clogged that it begins to impede the oil flow, a by-pass valve opens, and routes the oil flow through the filter core. This results in unfiltered oil being circulated throughout the engine, a condition which is avoided if the filter element is changed at regular intervals.

4   The oil pressure switch, which is situated at the top of the crankcase behind the cylinder block, serves to indicate when the oil pressure has dropped due to an oil pump malfunction, blockage in an oil passage, or a low oil content. The switch is not intended to be used as an indication of the correct oil level.

5   As previously mentioned an oil breather is incorporated into the system. It is mounted in the top of the camshaft cover and is essential for an engine of this size with so many moving parts. It serves to minimise crankcase pressure variations due to piston and crankshaft movement, and also helps lower the oil temperature, by venting the crankcase. Furthermore this system reduces the escape of unburnt oil into the atmosphere and so allows use of the machine in countries where stringent anti-pollution statutes are in operation. The breather tube carries the crankcase vapours to the air cleaner housing where they become mixed with the air drawn into the carburettors.

6   Excessive oil consumption indicated by blue smoke being emitted from the exhaust pipes, coupled with a poor performance and fouling of sparking plugs, is caused by either an excessive oil build-up in the oil breather chamber, or by oil

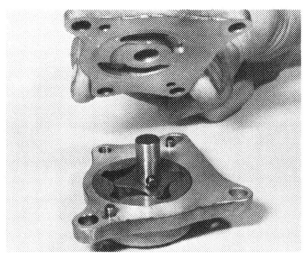

14.4 Separate the two halves of the oil pump

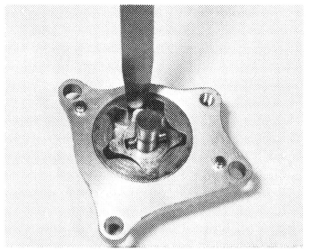

14.6 Check inner rotor to outer rotor clearance and ...

14.7 ... the rotor endfloat with a straight edge

getting past the piston rings. First check the oil breather chamber and air cleaner for oil build-up. If this is the fault, check the passageway from the air/oil separator in the oil breather chamber to the lower half of the crankcase. Blockage here will prevent oil flowing back into the crankcase, resulting in oil build-up in the breather chamber and air cleaner tube.

7 Be sure to check the oil level in the sump before starting the engine. If the oil level is not seen between the two marks adjacent to the sight window at the bottom of the clutch cover, replenish with the correct amount of oil of the specified SAE 10W/40 viscosity.

15 Oil pump: removal and examination

1 The oil pump is secured to the wall of the primary drive chamber behind the clutch unit. To gain access to the pump, the engine oil should be drained and the primary drive/ clutch cover detached. The clutch should then be removed as described in Chapter 1, Section 14.

2 Unscrew the three screws retaining the oil pump and lift it from position. Displace the two O-rings in the casing wall. The oil pump pinion is retained on the pump shaft. Remove the circlip, lift the pinion off the shaft and push out the drive pin.

3 Remove the single screw from the reverse side of the pump body. The two halves of the pump body are located by two tight

fitting dowel pins . Rather than levering the cases apart, which would damage the mating surfaces, the dowels should be driven out. Use a parallel shanked punch of a suitable size, whilst resting the pump across two strips of wood of a thickness sufficient to raise the pump off the workbench surface.

4 Separate the outer casing (reverse side) from the pump, leaving the drive shaft and rotors in place at this stage. Push out the drive shaft, together with the drive pin, and then lift out the two rotors.

5 Wash all the pump components with petrol and allow them to dry before carrying out a full examination. Before partly reassembling the pump for the various measurements to be made, check the castings for cracks or other damage, especially the pump end covers.

6 Reassemble the pump rotors and measure the clearance between the outer rotor and the pump body, using a feeler gauge. If the clearance exceeds 0.25 mm (0.0098 in) the rotor or the body must be renewed, whichever is worn. Measure the clearance between the outer rotor and the inner rotor with a feeler gauge. If this clearance is greater than 0.2 mm (0.008 in) the rotors must be renewed as a set.

7 Using a small sheet of plate glass or a straight edge placed across the pump housing, check the rotor endfloat. If the endfloat exceeds 0.15 mm (0.006 in) the complete pump must be renewed.

8 Examine the rotors and the pump body for signs of scoring, chipping or other surface damage which will occur if metallic particles find their way into the oil pump assembly. Renewal of the affected parts is the only remedy under these circumstances, bearing in mind that rotors must always be replaced as a matched set.

9 Reassemble the pump by reversing the dismantling procedure. Make sure all parts of the pump are well lubricated before the end cover is replaced and that there is plenty of oil between the inner and outer rotors. Apply a small quantity of locking fluid to the thread of the single casing screw. **Do not** omit the two O-rings when fitting the oil pump into the casing. Rotate the drive shaft as the screws are tightened down, to check that the oil pump revolves freely. A binding pump may be caused by dirt on the rotor faces or distortion of the cases, due to unequally tightened screws.

16 Checking the oil pressure

1 Because of the predominant use of caged ball and roller bearings in the GS series of engines, a low pressure lubrication system is employed. If the condition of the oil pump is suspect, the output pressure may be checked by connecting a suitable pressure gauge to the engine.

2   A blanking plug is fitted to the right-hand end of the main oil passage which runs across the crankcase below and to the rear of the cylinder block. The blanking plug should be substituted by a suitable adaptor piece to which the pressure gauge can be attached, via a flexible hose.

3   After connection of the pressure gauge, check that the oil level in the crankcase is correct and then start the engine. The engine should be run until the oil is at approximately 60°C (140°F). This temperature will be reached after approximately 10 minutes of running at 2000 rpm in the summer, and roughly twice as long at the same engine speed in winter. Because an air-cooled engine should not ideally be left to idle for 20 minutes without a cooling draught, an alternative would be to ride the machine for approximately 10 minutes with it operating around the 4000 rpm region. Raise the engine speed to 3000 rpm, when the pressure gauge should give a reading of 0.1 kg/cm^2 (1.42 psi). It can be seen that the pressure gauge must be of high sensitivity and of the correct calibration to give a useful reading. A pressure reading lower than specified may be caused by a worn oil pump or a blocked oil strainer or oil filter element. Before dismantling the pump for inspection clean the oil strainer and renew the oil filter, as described in Section 17 of this Chapter.

17.1 Oil filter chamber cover plate is retained by three domed nuts

16.4 Check switch by operating the plunger

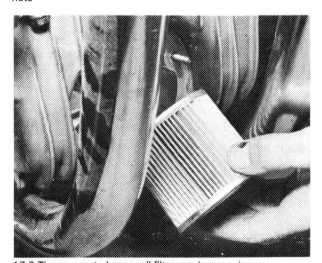

17.2 The corrugated paper oil filter requires regular replacement

17 Oil pressure warning switch

1   An oil pressure failure warning switch is screwed in to a holder bolted to the top of the crankcase. The switch is interconnected with a warning light in the instrument console.

2   If the oil warning lamp comes on whilst the machine is being ridden, the engine must be stopped immediately, otherwise there is risk of severe engine failure due to a breakdown of the lubrication system. The fault must be located and rectified before the engine is re-started and run even for a brief moment.

3   Oil pressure failure may be due to a blocked oil strainer screen or a blocked filter and by-pass valve. A worn oil pump or sheared drive shaft or pin will also produce the same symptoms.

4   Failure of the switch itself is also possible. This fault should be eliminated first before seeking the cause of pressure failure elsewhere. To check the switch, it should be removed from the crankcase by unscrewing the two bolts securing it and the holding cover to the crankcase. Switch on the ignition and check that the warning bulb has illuminated. Push the brass switch plunger upwards slowly. The lamp should go out. If the warning bulb remains lit the switch is faulty and should be renewed.

5   When the engine is operated at high temperatures, there may be a tendency for the oil warning lamp to come on occasionally, at idling speeds. This is quite in order if the light extinguishes immediately engine speed is increased.

18 Oil filter: renewing the element

1   The oil filter element is contained within a semi-isolated chamber in the front of the lower crankcase, closed by a finned cover retained by three domed nuts. Before removing the cover, place a receptacle below the engine to catch the engine oil contained within the filter chamber. The oil will drain from the chamber as the cover is released, with the three nuts removed. A coil spring is fitted between the cover and the filter element to keep the element firmly seated in position. Be prepared, therefore, for the cover to spring off after the retaining nuts have been removed.

2   No attempt should be made to clean the oil filter element; it must be renewed. When renewing the filter element it is wise to renew the filter cover O-ring at the same time. This will obviate the possibility of any oil leaks.

3   The by-pass valve which allows a continued flow of lubrication if the element becomes clogged is an integral part of the filter. For this reason routine cleaning of the valve is not required since it is renewed regularly.

4   Never run the engine without the filter element or increase the period between the recommended oil changes or oil filter changes. The oil should be changed every 2000 miles and the oil filter renewed at every second oil change.

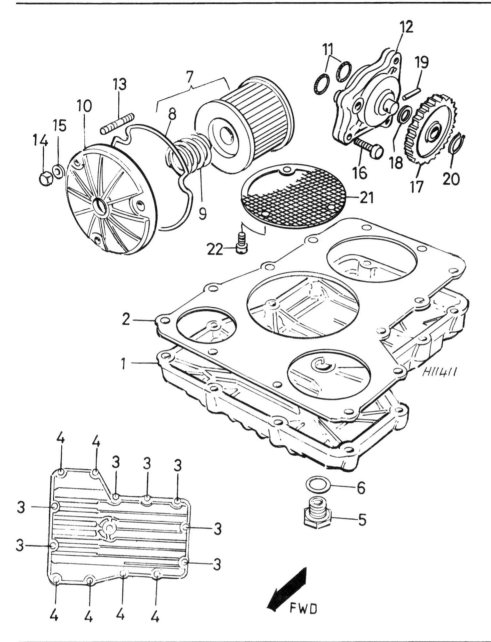

Fig. 2.8 Oil pump and filter

1   Sump
2   Sump gasket
3   Bolt – 7 off
4   Bolt – 6 off
5   Drain plug
6   Sealing washer
7   Oil filter assembly
8   O-ring
9   Spring
10 Filter cap
11 O-ring – 2 off
12 Oil pump
13 Stud – 3 off
14 Nut – 3 off
15 Washer – 3 off
16 Screw – 3 off
17 Oil pump pinion
18 Washer
19 Pin
20 Circlip
21 Oil strainer
22 Screw – 3 off

19 Fault diagnosis: fuel system and lubrication

Symptom	Cause	Remedy
Engine gradually fades and stops	Fuel starvation	Check vent hole in filler cap. Sediment in filter bowl or float chamber. Dismantle and clean.
Engine runs badly. Black smoke from exhausts	Carburettor flooding	Dismantle and clean carburettor. Check for punctured float or sticking float needle.
Engine lacks response and overheats	Weak mixture Air cleaner disconnected or hose split Modified silencer has upset carburation	Check for partial block in carburettors. Reconnect or renew hose. Replace with original design.
Oil pressure warning light comes on	Lubrication system failure	Stop engine immediately. Trace and rectify fault before re-starting.
Engine gets noisy	Failure to change engine oil when recommended	Drain off old oil and refill with new oil of correct grade. Renew oil filter element.

Chapter 3 Ignition system

Refer to Chapter 7 for information relating to the 1981 to 1988 models

Contents

Specifications

Ignition system	GS 850 GN	GS 850 GT/GLT
Type	Contact breaker	Fully transistorised
Ignition timing		
Retarded	17° BTDC below 1500 rpm	17° BTDC below 1500 rpm
Advanced	37° BTDC above 2350 – 2500 rpm	37° BTDC above 2350 – 2500 rpm
Firing order	1, 2, 4, 3	1, 2, 4, 3
Cylinder identification	Numbered consecutively 1 – 4, from alternator (left) end	
Contact breaker – GN model:		
Gap	0.3 – 0.4 mm (0.012 – 0.016 in)	
Dwell angle	180° (45° reading on 90° /4-cyl scale)	
Ignition coil		
Primary resistance (orange/white to white – cyl 1 and 4, orange/white to black or black/yellow – cyl 2 and 3)	About 4 ohms	About 4 ohms
Secondary resistance (plug cap to plug cap)	About 15 K ohms	About 32 K ohms
Minimum spark gap	8 mm	8 mm
Signal coil resistance	N/A	290 – 360 ohms
Condenser		
Capacity	0.16 – 0.20 microfarads	N/A
Sparking plugs		
Make	NGK or Nippondenso	
Type	B8ES or W24ES/W24ES-U	
Gap	0.6 – 0.8 mm (0.024 – 0.031 in)	

1 General description

There are two types of ignition system fitted to the various GS 850 models. The GN models utilise the traditional contact breaker points system, whereas the GT/GLT models are equipped with Suzuki's fully transistorised system. The operation of the two systems will be explained in turn, although it will be noted that certain stages of the ignition sequence are controlled in the same way despite the two differing methods of timing the final required spark.

On the GS 850 GN model, there are two sets of contact breaker points, two condensers, four sparking plugs and an automatic ignition advance mechanism. The contact breaker cam, which is incorporated in the advance mechanism, opens each set of points once in 180° of crankshaft rotation, causing a spark to occur in two of the cylinders. The other set of points fires 180° later, so that in every 360° of crankshaft rotation, each plug is fired once.

Each set of points has one fixed and one moveable contact, the latter of which pivots as the lobe of the cam separates them. The two condensers are wired in parallel, one with each set of contact points, and these function as electrical storage reservoirs, whilst also preventing arcing across the points. The condenser serves to absorb surplus current that tries to run back through the system when there is an overload, and feeds the current back to the ignition coils. They also help intensify the spark. When the points are closed, the current flows straight through them to earth. When they are open, there is now an open circuit. But for the condensers, the current would arc across the points causing them to burn and pit. When the condensers reach their capacity, they discharge the current back through the primary windings and eventually to the sparking plug.

On the GS 850 GT/GLT models, the ignition components are a signal generating unit, comprising two pick-up coils and a permanent magnet, a separate ignitor unit, two ignition coils, four sparking plugs, and an automatic ignition advance mechanism.

The signal generating device is located on the right-hand end of the crankshaft, in the position normally occupied by the contact breaker points, and behind a similar cover. The device transmits signals, in the form of electrical impulses, to the remotely mounted ignitor unit. This unit, which contains a transistor capable of amplifying each signal received, is fitted to the electrical components mounting plate, behind the left-hand side panel. As each signal is received in the ignitor unit, a transistor contained therein operates 'on' or 'off', and opens and closes the correct circuit as required. Thus, the correct impulse for the crankshaft position, that is, whichever pair of cylinders (Nos 1 and 4 or Nos 2 and 3) is firing, is received and passed on. By cutting off the primary current flowing on the primary side of the ignition coil, the sparking plugs are caused to spark. In other words, the primary current of the ignition coil is cut off by the transistor in this type of ignition system, whereas it is cut off by the contact breaker points in the conventional ignition system. The two ignition coils and the sparking plugs are the same items as fitted to the conventional type of ignition system.

There are several advantages that the transistorised ignition system holds over the traditional contact breaker points system. The most important advantage is that it removes the majority of mechanical components from the system. Because there are no contact breakers to wear, the owner is freed from the task of periodically adjusting or renewing them. Once the transistorised system has been set up, it need not be attended to unless it has been disturbed in the course of dismantling or failure occurs somewhere in the system. Other advantages of the transistorised ignition system include a greater resistance to the effects of vibration, dirt and moisture, and a constantly 'strong' spark at exactly the correct moment, with no wastage of electrical energy due to arcing etc.

On all the models, each of the two coils has two high voltage sparking plug leads, and one coil serves cylinders 1 and 4, and the other, cylinders 2 and 3.

The automatic advance mechanism serves to advance the ignition timing as the engine rpm rises. The mechanism is made up of two spring loaded weights which, under the action of centrifugal force created by the rotation of the crankshaft, fly apart and cause either the contact breaker points to open earlier, or the signal generator to release an impulse earlier. If the mechanism does not operate smoothly, the timing will not advance smoothly, or it will tend to stick in one position. This will result in poor running in any but that one position. Sometimes the springs are prone to stretching which can cause the timing to advance too soon. It is best to check the automatic advance mechanism, by carrying out a static timing test (850 GN models only), using the inscribed timing marks, followed by a strobe test.

The electrical system of all the models, is powered by an ac generator (alternator) fitted to the extreme left-hand end of the crankshaft. The alternating current (ac) is passed through a combined regulator and rectifier unit. The rectifier converts the ac to direct current (dc) and can then be used to charge the battery and provide current for the lights and ancillary components. Output of the alternator is controlled by the silicon-controlled regulator (SCR) unit to within a range of 14 – 15.5 volts with the engine operating at 5000 rpm.

2 Crankshaft alternator: checking the output

1 If the charging performance of the alternator is suspect, it can be checked with a multi-meter test instrument that includes a voltmeter and ohmmeter. As most owner/riders are unlikely to possess equipment of this type it is advised that the machine be returned to a Suzuki Service Agent for testing.
2 If a multi-meter is available, an initial check on the alternator and the rectifier and regulator assemblies may be carried out as described in Chapter 6. As mentioned in Chapter 6, Section 3 the charging system should be considered as a whole, and should be tested accordingly.

3 Ignition coils: checking

1 Each ignition coil is a sealed unit, designed to give long service without need for attention. They are located within the top frame tubes, immediately to the rear of the steering head assembly. If a weak spark and difficult starting causes the performance of a coil to be suspect, it should be tested by a Suzuki Service Agent or an auto-electrical engineer who will have the appropriate test equipment. A faulty coil must be renewed; it is not possible to effect a satisfactory repair.
2 On the GS 850 GN model, a defective condenser in the contact breaker circuit can give the illusion of a defective coil and for this reason it is advisable to investigate the condition of the condenser before condemning the ignition coil. Refer to Section 6 of this Chapter for the appropriate details.
3 Note that it is extremely unlikely that both ignition coils will prove faulty at the same time, unless the common electrical feed is in some way deranged. This can be checked by measuring the low tension voltage supplied to the coils, using a voltmeter.

4 Contact breaker: adjustments – GS 850 GN model only

1 To gain access to the contact breaker assembly, it is necessary to detach the aluminium cover retained by three crosshead screws at the right-hand end of the crankshaft. Note that the cover has a sealing gasket, to prevent the ingress of water.
2 Rotate the engine slowly by means of the engine turning hexagon until one set of points is in the fully open position. Examine the faces of the contacts. If they are blackened and burnt, or badly pitted, it will be necessary to remove them for further attention. See Section 5 of this Chapter. Repeat for the second set of contact points.
3 Adjustment is effected by slackening the screw through the plate of the fixed contact breaker point and moving the point either closer to or further from the moving contact until the gap is correct as measured by a feeler gauge. The correct gap with the points FULLY OPEN is 0.3 – 0.4 mm (0.012 – 0.016 in). Small projections on the contact breaker baseplate permit the insertion of a screwdriver to lever the adjustable point into its correct location. Repeat this operation for the second set of points, which must also be fully open.
4 Do NOT slacken the two screws through the extremities of the larger baseplate fitted to the right-hand set of contact breaker points. They are used for adjusting the setting of the ignition timing and it will be necessary to re-time the engine if the baseplate is permitted to move. Only the centre screw should be slackened, to adjust the fixed contact breaker point.
5 Before replacing the cover and gasket, place a light smear of grease on the contact breaker cam and one or two drops of thin oil on the felt which lubricates the surface of the cam. It is better to under-lubricate rather than add excess because there is always chance of excess oil reaching the contact breaker points and causing the ignition circuit to malfunction.

3.1 Each coil is secured by two bolts and serves two cylinders

4.3 Slacken single screw (arrowed) to adjust each set of points

5 Contact breaker points: removal, renovation and replacement – GS 850 GN model only

1 If the contact breaker points are burned, pitted or badly worn, they should be removed for dressing. If it is necessary to remove a substantial amount of material before the faces can be restored, the points should be renewed.

2 To remove the contact breaker points, detach the circlip which secures the moving contact to the pin on which it pivots. Remove the nut and bolt which secures the flexible lead wire to the end of contact return spring, noting the arrangement of the insulating washers so that they are replaced in their correct order during reassembly. Lift the moving contact off the pivot, away from the assembly.

3 The fixed contact is removed by unscrewing the screw which retains the contact to the contact breaker baseplate.

4 The points should be dressed with an oilstone or fine emery cloth. Keep them absolutely square throughout the dressing operation, otherwise they will make angular contact on reassembly, and rapidly burn away.

5 Replace the contacts by reversing the dismantling procedure, making sure that the insulating washers are fitted in the correct order. It is advantageous to apply a thin smear of grease to the pivot pin, prior to replacement of the moving contact arm.

6 Check, and if necessary, re-adjust the contact breaker gap when the points are fully open. Repeat the whole operation for the second set of points.

6 Condensers: removal and replacement – GS 850 GN model only

1 A condenser is included in each contact breaker circuit to prevent arcing across the contact breaker points as they separate. It is connected in parallel with each set of points and if a fault develops, ignition failure is liable to occur.

2 If the engine proves difficult to start, or misfiring occurs, it is possible that the condenser is at fault. To check, separate the contact breaker points by hand when the ignition is switched on. If a spark occurs across the points and they have a blackened and burnt appearance, the condenser can be regarded as unserviceable.

3 It is not possible to check a condenser without the appropriate test equipment. In view of the low cost involved, it is preferable to fit a new one and observe the effect on engine performance.

4 Because each condenser and its associated set of contact breaker points is common to a pair of cylinders, a faulty condenser will not cause a misfire on one cylinder only. In such a case, it is necessary to seek the cause of the trouble elsewhere, possibly in some other part of the ignition circuit or the carburettors. It also follows that both condensers are unlikely to fail at the same time unless damaged in an accident. If the cases are crushed or dented, electrical breakdown will occur.

5 The condensers are located at the base of the contact breaker assembly, parallel to each other. Each has an integral bracket and is attached to the contact breaker baseplate by a single crosshead screw, making renewal simple.

6.5 Condensers are held by a single screw each below the points assembly

7 Ignition timing: checking and resetting

GN model only

1 In order to check the accuracy of the ignition timing, it is necessary to remove the contact breaker cover from the right-hand side of the crankcase. Ignition timing checking and resetting should take place after resetting the contact breaker gaps, as described in Section 4.

Fig. 3.1 Contact breaker assembly – GN model

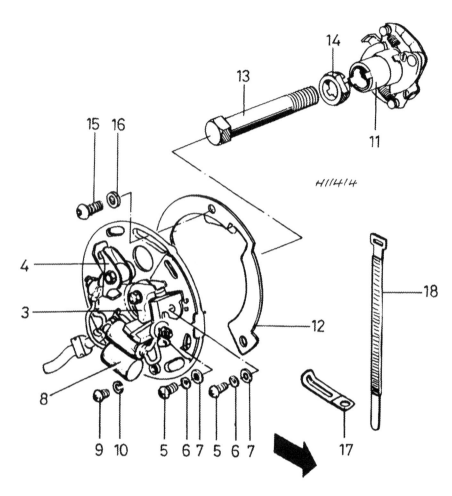

1   Contact breaker assembly
2   Base plate
3   Right-hand contact breaker
4   Left-hand contact breaker
5   Screw – 4 off
6   Spring washer – 4 off
7   Washer – 4 off
8   Capacitor – 2 off
9   Screw – 2 off
10 Spring washer – 2 off
11 Automatic timing unit
12 Timing plate
13 Bolt
14 Engine turning hexagon
15 Screw – 3 off
16 Washer – 3 off
17 Cable clip
18 Cable clip

2   Apply a spanner to the engine rotation hexagon and turn the engine in a forward direction, whilst viewing the ATU through the inspection aperture in the contact breaker stator plate. It will be seen that there is a set of three scribed lines on each side of the ATU.

3   Commence ignition timing checking on the left-hand contact breaker set, which controls cylinders No 1 and 4. To determine at which moment the points open connect a 12v bulb between the moving point and a suitable earthing point on the engine. With the ignition turned on, the bulb will light up when the points are open. Rotate the engine until the F1-4 mark on the ATU is in **exact** alignment with the index pointer mark on the plate fitted to the rear of the stator plate. If the ignition is correct, the points should be on the verge of opening when this position is reached. This will be indicated by the flickering of the bulb.

4   To adjust the ignition timing on No 1 and 4 cylinders slacken the three screws which pass through the elongated holes in the stator plate periphery. Rotate the plate until the light flickers and then tighten the screws. Turn the engine backwards about 90° and then forwards again to check the setting.

5   Check the ignition timing on Nos 2 and 3 cylinders in a similar manner, using the F2-3 timing mark. If the timing is incorrect, slacken the two screws holding the right-hand con-

tact breaker assembly plate to the main stator plate. Move the plate to the correct position and tighten the screws. Recheck the timing.

6   Provided that the contact breakers are in good condition and care is taken, manual adjustment of the ignition timing should be acceptably accurate. If possible, however, the timing should be checked using a stroboscope lamp because not only can the accuracy of the timing be checked with the engine running but the correct performance of the ATU can be verified. The timing light should be connected to the low tension or high tension side of the ignition as instructed by the manufacturers of the light. Test the left-hand contact breaker and then the right-hand contact breaker. Start the engine and illuminate the ATU through the inspection aperture. With the engine running below 1500 rpm the F mark should be in alignment with the index mark. Raise the engine speed slowly to 2500 rpm when the advance mark should align with the index pointer.

7   The advance range of 1000 rpm peaks at about 2500 rpm, above which engine speed no more advance is possible. If, when increasing the engine speed from the commencement of advance at 1500 rpm, the timing marks are seen to move erratically, or if the advance range has altered appreciably, the ATU should be inspected for wear or malfunctioning as described in the following Section.

GT/GLT models only

8   Although the ignition system fitted to these models appears to be of the fixed, non-adjustable, type, in practice a small amount of adjustment is possible. This adjustment facility may be of benefit in one or two ways. If the machine is to be fitted with, or has been recently fitted with, a new signal generating unit, the positioning of the device can be checked, and if necessary, adjusted. Also if the machine is malfunctioning, and the case of the problem cannot be traced to any other source, an investigative check of the ignition timing may prove beneficial.

9   In order to check the accuracy of the ignition timing on these models, it is first necessary to remove the small circular cover from the right-hand side of the crankcase. The ignition timing on this system can only be checked using a stroboscope timing lamp. In this way an additional task, that of checking the correct function of the ATU, may be accomplished simultaneously with checking the ignition timing. Connect the timing light to the low tension or high tension side of the ignition as instructed by the manufacturers of the light. Start the engine and illuminate the ATU through the inspection aperture. With the engine running below 1500 rpm ($\pm$ 150 rpm), the F mark should be alignment with the index mark. Raise the engine speed slowly to 2350 rpm ($\pm$ 150 rpm) when the advance mark should be in alignment with the index pointer.

10   If the ignition timing is slightly incorrect, the three screws holding the signal generator mounting plate should be slackened. Move the mounting plate, the screw holes are slightly elongated to permit this movement, and recheck the ignition timing. Slight movements may be made until the timing is correct.

11   To ensure the ATU is functioning correctly, refer to paragraph 7 of this Section.

7.4 Use three screws in elongated holes in stator plate edge, to adjust cylinders 1 and 4 timing (one shown, arrowed)

---

8   Automatic timing unit: examination – all models

1   The automatic timing unit rarely requires attention although it is advisable to examine it periodically.

2   To obtain access to the unit remove the inspection cover and the contact breaker back plate complete with contact breakers. The ATU centre bolt and engine turning hexagon should be removed before the stator plate. Before removal, the back plate and end cover should be marked so that the back plate can be replaced in exactly the same position. This will ensure the ignition timing is not altered.

3   Pull the ATU from position, noting the drive pin with which it locates and is driven. The unit comprises balance weights which move outwards against spring tension as the centrifugal forces increase. The balance weights must move freely on their

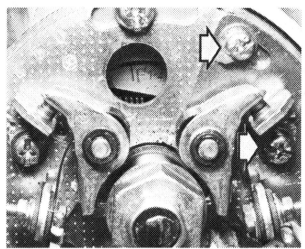

7.3 Timing index marks for cylinders 1 and 4 are visible. Timing adjustment screws for other pair (2 and 3) are arrowed

pivots, which should be lubricated. The tension springs must also be in good condition.

4   In addition, on the GN model, check the surface of the contact breaker cam for pitting or obvious signs of wear. Damage to the cam cannot be rectified; the complete ATU must be renewed.

5   When replacing the ATU, check that the drive pin engages with the recess in the rear of the centre boss. Because there is a single recess only, the ATU cannot inadvertently be replaced in the incorrect position and so alter the timing marks in relation to the crankshaft.

---

9   Transistorised ignition system: method of operation – GS 850 GT/GLT models

---

1   Signal current for this type of ignition system is provided by the two pick-up coils attached to a mounting plate, in a manner similar to the two sets of contact breakers on the GS 850 GN model. The two coils are fitted to the mounting plate, one either side of the protruding head of the ATU centre bolt and engine turning hexagon. The other main component of this, the ignition signal generator, is the iron rotor. This is attached to the ATU. Each of the two pick-up coils contains a magnet at their base. As the rotor trip is rotated past the magnets of the pick-up coils, at 180° apart on the mounting plate, alternating current (ac) is produced which is then transmitted to the ignitor unit.

2   The ignitor unit, mounted on the electrical components mounting plate and fitted behind the left-hand side panel of the machine, contains the system's transistors. The functions of the transistors fall under four main headings: 1 amplification; 2 switching; 3 oscillation, and 4 modulation. Each transistor has three terminals; these can be identified as the Base (B), Collector (C), and Emitter (E). They operate in the manner described in the following paragraph.

3   On the type of transistors used in this system, these being known as NPN type (Negative/Positive/Negative), the base (B) is the controlling terminal of the transistor operation. With the NPN type, the base utilizes only a positive or incoming signal in order to operate an 'ON' or 'OFF' switching function. The collector (C) is the terminal where voltage is accepted into the transistor. The emitter (E) is the terminal which performs the 'passing-on' function, of allowing the current through for use by the ignition coils, when the base terminal has received the correct signal impulse. Usually the voltage passed across the collector terminal to the emitter terminal is much larger than is needed at the base. This allows a relatively low voltage at the base terminal to control large working voltages across the collector to the emitter.

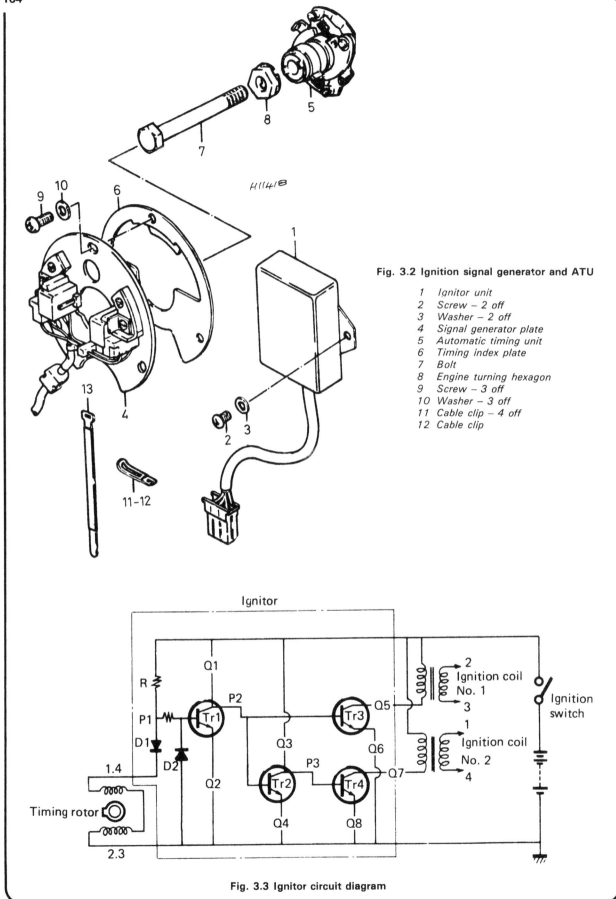

H11418

Fig. 3.2 Ignition signal generator and ATU

1   Ignitor unit
2   Screw – 2 off
3   Washer – 2 off
4   Signal generator plate
5   Automatic timing unit
6   Timing index plate
7   Bolt
8   Engine turning hexagon
9   Screw – 3 off
10 Washer – 3 off
11 Cable clip – 4 off
12 Cable clip

Ignitor

Ignition coil No. 1

Ignition coil No. 2

Ignition switch

Timing rotor

Fig. 3.3 Ignitor circuit diagram

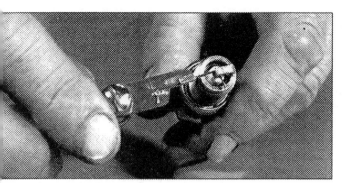

Electrode gap check - use a wire type gauge for best results

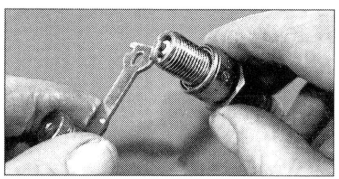

Electrode gap adjustment - bend the side electrode using the correct tool

Normal condition - A brown, tan or grey firing end indicates that the engine is in good condition and that the plug type is correct

Ash deposits - Light brown deposits encrusted on the electrodes and insulator, leading to misfire and hesitation. Caused by excessive amounts of oil in the combustion chamber or poor quality fuel/oil

Carbon fouling - Dry, black sooty deposits leading to misfire and weak spark. Caused by an over-rich fuel/air mixture, faulty choke operation or blocked air filter

Oil fouling - Wet oily deposits leading to misfire and weak spark. Caused by oil leakage past piston rings or valve guides (4-stroke engine), or excess lubricant (2-stroke engine)

Overheating - A blistered white insulator and glazed electrodes. Caused by ignition system fault, incorrect fuel, or cooling system fault

Worn plug - Worn electrodes will cause poor starting in damp or cold weather and will also waste fuel

10 Transistorised ignition system: testing and maintenance – GS 850 GT/GLT models

1   As stated earlier in this Chapter, the transistorised ignition system needs no regular maintenance once it has been set up and timed accurately. Occasional attention should, however, be directed at the various connections in the system, and these must be kept clean and secure. A failure in the ignition system is comparatively rare, and usually results in a complete loss of ignition. Usually, this will be traced to the ignitor unit, and little can be done at the roadside to effect a repair. In the event of the ignitor failing, it must be renewed as a repair is not practicable.

2   If the ignitor unit is thought to be at fault, it is recommended that it be removed and taken to a Suzuki dealer for testing. Testing at home is less practical and cannot be guaranteed to be accurate. At best, it will enable the owner to establish which part of the system is at fault, although replacement of the defective part remains the only effective cure.

3   For those owners possessing a multimeter and who are fully conversant with its use, a test sequence is given below for the ignitor unit and the signal generator. It is not recommended that the inexperienced attempt to test the system at home, as more damage could be sustained by the system.

11 Ignitor unit: test sequence – GS 850 GT/GLT models

1   Remove the left-hand side frame cover to expose the mounting plate on which the main electrical components are mounted.

2   Unscrew and remove the sparking plugs from cylinders No 1 and 2. Fit each of the sparking plugs to their respective plug caps and rest the caps and plugs on the top of the uppermost cylinder head fins. Disconnect the small two-pin connector block at the top, left-hand side of the electrical component mounting plate, in order to separate the leads from the signal generating unit. Turn the ignition switch, at the handlebars centre, to the 'ON' position.

3   Connect the positive probe of the multimeter, with the multimeter set to the X1 ohms scale, with the Blue lead of the two from the signal generating unit. Similarly, connect the negative probe of the multimeter to the Green lead from the signal generating device. When connecting the two probes, ensure they are connected with the ignitor unit section of the connector block; the half of the block nearest the ignitor unit. The ignitor unit is functioning correctly if the following is observed. The moment that the test probes are connected the sparking plug from No 1 cylinder sparks, and the moment the test probes are disconnected from the leads, No 2 cylinder sparking plug sparks. It should be noted that this test is carried out with the assumption that the ignition coils are in correct working order. It follows, therefore, that a faulty ignition coil will adversely effect the results of this test. See Section 3 of this Chapter for relevant information on the ignition coils.

12 Signal generating unit: test sequence – GS 850 GT/GLT models

1   Remove the left-hand frame cover from the machine. Set the multimeter to the X100 ohms range. Connect one multimeter probe to the Blue lead from the signal generator, at the connector block at the top, left-hand corner of the electrical components mounting plate. Connect the second probe to the Green lead at the block connector.

2   The resistance between the two lead wires should be between the range 250 – 360 ohms. If the indicated resistance falls below the lower figure or the meter gives a reading of infinite resistance, then the signal generating unit is in need of replacement. Before discarding the unit, however, a second opinion should be sought from a Suzuki Service Agent to ensure your diagnosis is indeed correct. The signal generating unit can only be purchased and replaced as a complete unit; the individual pick-up coils cannot be renewed individually.

13 Sparking plugs: checking and resetting the gaps

1   All the GS 850 models are fitted with either NGK or Nippon Denso (ND) sparking plugs. The standard types are NGK B-8ES or ND W24ES/W24ES-U, gapped within the range 0.6 – 0.8 mm (0.024 – 0.031 in). Operating conditions may indicate a change in sparking plug grade; the type recommended by the manufacturer gives the best, all round service.

2   Check the gap of the plug points every four monthly or 4000 mile service. To reset the gap, bend the outer electrode to bring it closer to the centre electrode and check that a 0.6 mm (0.024 in) feeler gauge can be inserted. Never bend the central electrode or the insulator will crack, causing engine damage if the particles fall in whilst the engine is running.

3   With some experience, the condition of the sparking plug electrodes and insulator can be used as a reliable guide to engine operating conditions.

4   Beware of overtightening the sparking plugs, otherwise there is risk of stripping the threads from the aluminium alloy cylinder heads. The plugs should be sufficiently tight to sit firmly on their copper sealing washers, and no more. Use a spanner which is a good fit to prevent the spanner from slipping and breaking the insulator.

5   If the threads in the cylinder head strip as a result of over tightening the sparking plugs, it is possible to reclaim the head by the use of a Helicoil thread insert. This is a cheap and convenient method of replacing the threads; most motorcycle dealers operate a service of this kind.

6   Make sure the plug insulating caps are a good fit and have their rubber seals. They should also be kept clean to prevent tracking. These caps contain the suppressors that eliminate both radio and TV interference.

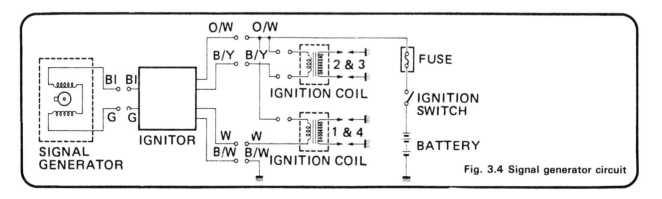

Fig. 3.4 Signal generator circuit

14 Fault diagnosis: ignition system

Symptom	Cause	Remedy
Engine will not start	Faulty ignition switch	Operate switch several times in case contacts are dirty. If lights and other electrics function. switch may need renewal.
	Starter motor not working	Discharged battery. Use kickstart until battery is recharged (GN model). Other models – find steep hill.
	Short circuit in wiring	Check whether fuse is intact. Eliminate fault before switching on again.
	Completely discharged battery	If lights do not work, remove battery and recharge.
Engine misfires	Faulty condenser in ignition circuit (GN model)	Renew condenser and re-test (GN model).
	Fouled sparking plug	Renew plug and have original cleaned.
	Poor spark due to generator failure and discharged battery	Check output from generator. Remove and recharge battery.
Engine lacks power and overheats	Retarded ignition timing (GN model)	Check timing and also contact breaker gap (GN model) Check whether auto-advance mechanism has jammed
Engine 'fades' when under load	Pre-ignition	Check grade of plugs fitted; use recommended grades only. Check ignition timing.

Chapter 4 Frame and forks

Refer to Chapter 7 for information relating to the 1981 to 1988 models

Contents

Specifications

Frame
Type .... Tubular steel, double cradle

Front forks
Type .... Telescopic, coil sprung hydraulically damped
Adjustment:
    UK models .... 3-way adjustable spring preload
    US models .... Air assisted, separate valves
Wheel travel .... 160 mm (6.3 in)

	GS850 GN, GT (UK)	GS850 GN, GT (US)	GS850 GLT
Fork spring free length:			
Standard	475.0 mm (18.7008 in)	421.0 mm (16.5748 in)	N/Av
Minimum	462.5 mm (18.2086 in)	416.0 mm (16.3779 in)	426.0 mm (16.7716 in)
Fork oil quantity – per leg	213 cc (7.50 Imp fl oz)	251 cc (8.49 US fl oz)	302 cc (10.21 US fl oz)
Fork oil type	SAE 10W/20	SAE 10W/20	SAE 10W/20

Fork oil level – measured from top of oil to top of stanchion, leg removed from machine and held upright, top bolt or plug, spacer(s), and fork spring removed:

Leg fully compressed	203 mm (7.99 in)	140 mm (5.51 in)	160 mm (6.30 in)
Leg fully extended	460 mm (18.11 in)	N/Av	N/Av

Fork air pressure – US models only:

	GS850 GN, GT	GS850 GLT
Standard pressure	8.5 psi (0.6 kg/cm^2)	11.4 psi (0.8 kg/cm^2)
Adjustment range	8.5 – 17 psi (0.6 – 1.2 kg/cm^2)	N/App
Maximum difference between legs	1.4 psi (0.1 kg/cm^2)	1.4 psi (0.1 kg/cm^2)
Maximum operating pressure	17 psi (1.2 kg/cm^2)	N/App
Maximum absolute pressure	35.5 psi (2.5 kg/cm^2)	35.5 psi (2.5 kg/cm^2)

Note: Air pressure is measured with the forks cold, machine on its centre stand with the front end jacked up so that all weight is removed from the (fully-extended) forks

Rear suspension
Type .... Swinging arm, welded tubular steel
Suspension units .... Hydraulically damped, 5-way adjustable spring preload, and
                                               4-way adjustable damping.
Rear wheel travel .... 100 mm (3.9 in)

Main torque wrench settings

Steering head nut	3.5 – 5.0 kgf m (25.3 – 36.0 lbf ft)
Upper yoke pinch bolt	2.0 – 3.0 kgf m (14.5 – 21.5 lbf ft)
Lower yoke pinch bolt	1.5 – 2.5 kgf m (11.0 – 18.0 lbf ft)
Front footrest mounting bolts	2.7 – 4.3 kgf m (19.5 – 31.0 lbf ft)
Swinging arm pivot nut	11.0 – 13.0 kgf m (79.5 – 94.0 lbf ft)
Rear suspension units:	
Upper mounting nuts	2.0 – 3.0 kgf m (14.5 – 21.5 lbf ft)
Lower mounting nuts	2.0 – 3.0 kgf m (14.5 – 21.5 lbf ft)

1 General description

The Suzuki GS 850 models all share the same conventional duplex cradle frame. This type of frame has the engine suspended inside the frame tubes rather than the engine constituting any part of the frame.

The suspension on these models does not follow convention to the same extent as the frame. The front forks, whilst being of the usual telescopic type, have unusual internal damping arrangements. The damping medium is still predominently oil, but on the GS 850 models sold on the US market, the forks have the added benefit of variable air assistance. Conventional fork springs are also contained within each fork stanchion, and it is the springs that provide some variation in fork action on the UK specification models. The use of air enables the damping characteristics of the front forks to be adjusted almost infinitely over the prescribed range, to suit particular riders and riding conditions. This adjustment is simply made by increasing or reducing the pressure contained within each fork leg. The UK versions of the GS 850 are not equipped with the air assisted forks, but instead are fitted with a fork spring preload adjustment device. This device offers the rider the facility of three-way adjustment of the fork spring tension rate, to cope with differing riding conditions.

In addition to the above, the GS 850 GLT (Low Slinger) model is fitted with variable pressure front forks that carry the front spindle forward of the legs. This fork design, usually referred to as leading-axle, appears to provide extra travel and so conform to the overall styling image presented in the 'L' model.

Air is added to the front forks of the models thus equipped, by means of a Shraeder valve, enclosed beneath a chrome protector cap, at the top of each fork leg.

Rear suspension is provided by a swinging arm fork, pivoting on two taper roller bearings located by adjustable stubs, and supported by two fluid filled suspension units. These rear suspension units also offer more capacity for fine tuning than conventional units. Each unit incorporates 5-way adjustment of the spring preload, and a further 4-way rebound damping adjustment facility. The latter adjustment is facilitated by the provision of a knurled ring setting wheel at the top of each rear unit below a protective rubber collar. A total of nine different combinations of spring preload and damping force are available to cope with all riding requirements. The left-hand longitudinal swinging arm member also serves as a torque tube through which the final driveshaft passes, and to which is attached the final drive gear casing.

2 Front forks: removal from the frame

1 It is unlikely that the front forks will need to be removed from the frame as a complete unit, unless the steering head bearings require attention or the forks are damaged in an accident. In the event of damage to one or both of the fork legs, they may be removed from the steering head and lower fork yoke, whilst leaving the two latter components in situ on the frame. Follow the procedure from paragraphs 5 to 9 if this approach is required.

2 Commence complete fork removal operations by removing the control cables from the handlebar levers or by removing the levers complete with cables. The shape of the handlebars fitted and the length of control cables will probably dictate the method used. Remove the front brake lever master cylinder unit, which is retained by a clamp held by two bolts. On the GS 850 GN model, ensure that the screw-on fluid reservoir cap is secure. Similarly, on the GT/GLT models, ensure that the four screws retaining the fluid reservoir cover are tight, before detaching the assembly from the handlebars. Tie the master cylinder to some part of the machine not to be dismantled, so that it is secure and resting in an upright position, and so that the weight is not taken by the brake hose. It is quite possible to remove the forks without having to separate the brake components from one another. As this method eliminates the necessity of bleeding the front brake on reassembly, it is herein described.

3 Detach the handlebars from their mounting points on the fork upper yoke. The handlebars are held by two U-clamps, retained by two bolts and spring washers each. In order to gain access to the handlebar mounting clamps and bolts, it is necessary to detach the separate plastic housing (not GLT model) fitted over the centre of the handlebars, through which the choke operating knob protrudes. Detach the housing after removing the four crosshead retaining screws. There is no need to detach the choke knob at this juncture; the housing will lift off over the knob. Remove the headlamp unit from the headlamp shell and disconnect the electrical leads at the snap connectors. No difficulty should be encountered in replacement, as the connections are mainly of the block type and the wiring is colour coded.

4 Disconnect the speedometer and tachometer cables at the instrument, where they are retained by knurled rings. Detach the wiring connections from the instrument bulb holders and the warning lamp console and from the gear position indicator unit and the fuel gauge, at their respective block and snap connectors. Undo the two bolts which pass through the instrument mounting bracket and lift the complete unit from the machine.

5 If the machine is not already resting on the centre stand, support it in this manner on firm, level grund. Balance the machine so that the front wheel is clear of the ground and place some robust packing, or ideally a jack, under the crankcase, so that if the machine should inadvertently tip forward, it will not roll off the centre stand.

6 Remove the front wheel as described in Chapter 5, Section 3. To permit front wheel removal, one caliper unit must be detached from the fork leg to which it is secured. In this case, however, both units should be detached, to allow subsequent fork leg removal. Remove the two bolts which secure each caliper, and temporarily, tie each caliper to the frame down tubes so that the weight is not taken by the hydraulic hoses.

7 Remove the front mudguard, which is retained by two bolts and lock washers passing into the inside of each fork leg.

8 On US specification models, remove the chromed valve protector caps from the top of each fork leg. Release all the air pressure from the forks by depressing the centre of the Shraeder valve.

9 Loosen the clamp bolts which retain the fork legs in the upper and lower yokes. The fork legs can now be eased downwards, out of position. If the clamps prove to be excessively tight, they may be gently sprung, using a large screwdriver. This must be done with great care, in order to prevent breakage of the clamps, necessitating renewal of the complete yoke.

Alternatively, if the fork legs appear to be binding in their clamps, such as may occur if they are bent or twisted as the result of an accident, liberal use of a solution of warm water and washing-up liquid, applied to the rubber seals in the clamps, can be very beneficial.

10 Remove the single bolt which secures the hydraulic hose union to the fork lower yoke. The complete front brake assembly, including the master cylinder, hoses and calipers, may be lifted away from the machine without the need to drain the fluid. If required, the lower yoke/hydraulic hose union cover plate may be detached at this stage.

11 Loosen the clamp bolt located at the rear of the centre of the upper yoke. Before the large chromed top bolt can be removed, the choke operating knob must be detached. Remove the petrol tank (see Chapter 2, Section 2) and then disconnect the choke cable from its holder and adjuster bracket. Slacken the choke knob locknut, turn the knob anti-clockwise and lift the knob away from the top of the steering head stem complete with its actuating cable. When the cable reaches its clamp below the front of the left-hand ignition coil, the clamp should be released to allow the cable to be pulled through. On the GLT model, detach the plastic cable shroud to the rear of the steering stem on the left-hand side, to facilitate removal of the choke cable from its clamp. Remove the large chromed bolt from the centre of the top yoke, together with the heavy washer fitted below the nut. From the underside, tap the upper yoke upwards until it frees the steering column. Support the weight of the lower yoke and, using a C-spanner, remove the steering head bearing adjuster ring. If a C-spanner is not available, a soft brass drift and hammer may be used to slacken the nut. With the top yoke detached, and whilst still supporting the weight of the steering stem and lower yoke, the complete assembly comprising the fork shrouds, the headlamp shell and the indicator lamps and their stems, can be lifted clear.

12 Remove the dust excluder and outer race (cone) once the adjuster nut has been detached. The bottom yoke, complete with steering column can now be lowered from position. Unlike many present-day motor cycles, the GS 850 models are fitted with taper roller bearings in the steering head. The lower bearing inner race will be lowered from position as the steering column is removed, leaving the outer race in position. Separate the bearing from the column, noting the positioning of a heavy washer below the lower race. The complete upper tapered roller bearing will remain in the top of the head lug casting.

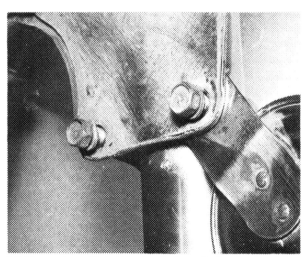

2.7 The mudguard is held by two bolts on the inside of each leg

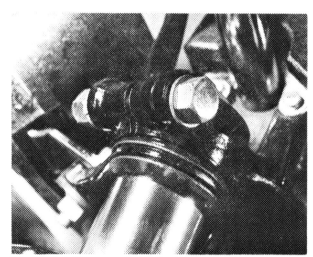

2.9a Slacken the upper pinch bolts (two) and ...

2.9b ... then the lower pinch bolts (four) and ...

2.9c ... withdraw each leg individually

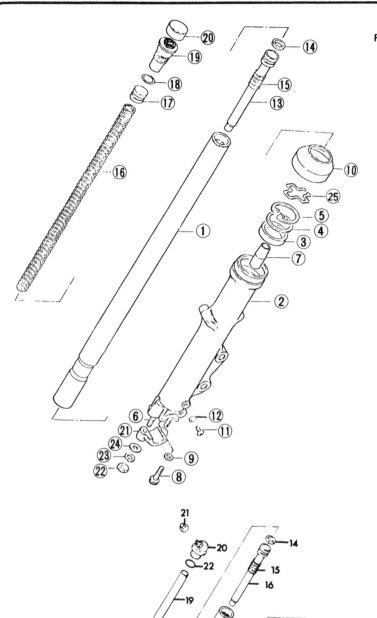

Fig. 4.1 Front forks – GS 850 GN and
GT model (UK)

1   Stanchion
2   Lower leg
3   Oil seal
4   Washer – GN model
5   Circlip – GN model
6   Stud – 2 off
7   Damper rod seat
8   Allen bolt
9   Sealing washer
10 Dust seal
11 Drain screw
12 Sealing washer
13 Damper rod
14 Piston ring
15 Rebound spring
16 Spring
17 Free piston
18 O-ring
19 Top bolt
20 Dust cap
21 Wheel spindle clamp
22 Nut – 2 off
23 Spring washer – 2 off
24 Washer – 2 off
25 Circlip – GT model

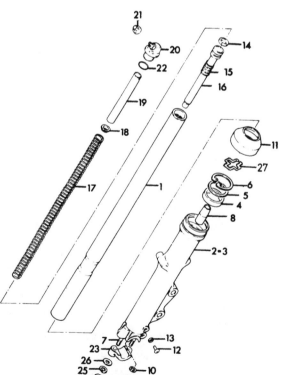

Fig. 4.2 Front forks – US models

1   Stanchion
2   Right-hand lower leg
3   Left-hand lower leg
4   Oil seal
5   Washer – GN model
6   Circlip – GN model
7   Stud – 2 off
8   Damper rod seat
9   Allen bolt
10 Sealing washer
11 Dust excluder
12 Drain screw
13 Sealing washer
14 Piston ring
15 Rebound spring
16 Damper rod
17 Spring
18 Seat
19 Spacer
20 Top bolt/valve
21 Valve cap
22 O-ring
23 Clamp*
24 Nut – 2 off*
25 Spring washer – 2 off*
26 Washer – 2 off*
27 Circlip – GT, GLT models

*GN and GT models only

3   Front forks: dismantling

1   It is advisable to dismantle each fork leg separately, using an identical procedure. There is less chance of unwittingly exchanging parts if this approach is adopted. Commence by draining each fork leg of damping oil; there is a drain plug in each lower leg above and to the rear of the wheel spindle housing. On the US market models, unscrew and remove the valve and then unscrew and remove the valve cap unit, noting the O-ring below the valve cap unit. On non air-assisted forks, prise off the rubber fork top cap, unscrew and remove carefully the spring preload adjuster piece. From the air-assisted forks, withdraw one long spring, a spring guide and a spacer. From the UK type forks, remove a spring compression piece and one long spring.

2   Clamp the fork lower leg in a vice fitted with soft jaws, or wrap a length of rubber inner tube around the leg to prevent damage. Unscrew the socket screw, recessed into the housing which carries the front wheel spindle, on all the models except the GLT, where the socket screw is right at the bottom of the leading-axle fork legs. In order to remove the socket screw successfully, the damper rod into which seat the socket screw is fitted, must be prevented from turning. Construct a simple device (see accompanying photograph), consisting of a nut or bolt of the correct size (19 mm) inserted and secured in the end of a length of suitable tubing. Alternatively, a bolt with a head of the correct size and two nuts fitted to the end of an extension bar and socket will suffice. Whichever device is used, insert the locking nut or bolt into the damper rod upper end and unscrew the socket screw with the correct sized Allen key. Prise the dust excluder from position and slide it up the fork upper tube. The upper tube (stanchion) can be pulled out of the lower fork leg. Pull the damper rod seat off the rod, invert the upper tube and push the damper rod out of position towards the top end of the tube.

3   The oil seal fitted to the top of the lower leg should be removed only if it is to be renewed. Renewal of the oil seals is the only course of action if their ability to contain the fork oil has been impaired. With the air-assisted forks fitted to certain models in the GS 850 range, a secondary function of oil seals has to be considered. They must be capable of containing the required air pressure, without any pressure losses. This is particularly important in view of the small amount of pressure involved in the normal operation of the forks. It follows that with only a small quantity of air pressurized, it only needs a slight leakage to seriously affect the amount of air contained within the forks. It is unlikely that the seals would fail simultaneously and therefore an imbalance would occur in damping which could seriously affect handling. The reason for not disturbing an oil seal unless failure has occurred is that damage will almost certainly be inflicted when it is prised from position. The seal is retained by a washer and circlip on the GN model, and by a shaped stopper ring on the GT/GLT models.

4   Note the copper sealing washer fitted to the socket head bolt at the bottom of the fork legs. Ensure it is not misplaced, and is reinstalled upon reassembly.

4   Front forks: examination and renovation

1   The front forks do not contain bushes. The fork legs slide directly against the hard chrome surface of the fork tubes. The parts most likely to wear, therefore, over an extended period of service are the wearing surfaces of the fork stanchions (tubes) and lower legs. If wear occurs here, indicated by slackness, the fork leg complete will have to be renewed, possibly also the fork stanchion. Wear on the fork stanchion is indicated by scuffing and penetration of the hard chrome surface. In addition, wear can occur in the damper assembly within the fork stanchion and the oil seal at the sliding joint. Wear is normally accompanied by a tendency for the forks to judder when the front brake is applied, and it should be possible to detect the increased amount of play by pulling and pushing on the handlebars when the front brake is applied fully. This type of wear should not be confused with slack steering head bearings, which can give identical results.

2   Renewal of the worn parts is quite straightforward. Particular care is, however, necessary when renewing the oil seal. Ensure the edges of the top of the lower leg do not become damaged during the operation to prise out the oil seal. Both the seal and the fork stanchion should be lightly greased upon reassembly, to lessen the risk of damage to the new seal as it is fitted.

3   After an extended period of service, the fork springs may take a permanent set. The service limit for the fork springs, when they are not under compression, in other words when they are free, is shown in the specifications at the beginning of the Chapter, for the different models. Always fit new springs as a pair, **never** separately.

4   Check the outer surface of the stanchion for scratches or roughness. It is only too easy to damage the oil seal during reassembly, if these high spots are not eased down. The stanchions are unlikely to bend unless the machine is damaged in an accident. Any significant bend will be detected by eye, but if there is any doubt about straightness, roll the stanchion tubes on a flat surface. If the stanchions are bent, they must be renewed. Unless specialised repair equipment is available, it is rarely practicable to effect a satisfactory repair to a damaged stanchion.

5   The piston ring fitted to the damper rod may wear if oil changes at the specified intervals are neglected. If damping has become weakened and does not improve as a result of an oil change, the piston ring should be renewed. Check also that the oilways in the damper rod have not become obstructed.

5   Steering head bearings: examination and renovation

1   Clean off any dirt that has accumulated on the upper and lower bearing races. Check the bearing races thoroughly for any signs of wear or damage. No signs of indentation or pitting, and a smooth polished appearance should be apparent. Renew the set as a whole if necessary.

2   Clean and examine the bearing rollers. Check the rollers for any signs of damage, such as scoring marks, and check for a uniform polished appearance. If any require replacement the whole set must be renewed.

3   The outer races are a drive fit in the steering head lug and may be drifted out, using a suitable long handled drift passed through the centre of the lug. The lower inner race may be levered from position on the steering stem. When driving the new inner races into place, ensure that they remain square to

3.1a Remove drain plugs to drain fork oil from legs

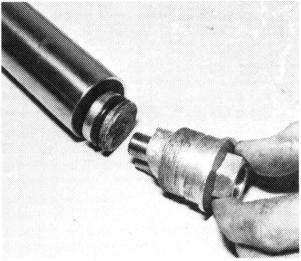

3.1b Unscrew the spring preload adjuster piece to free the spring compression piece etc (UK models). Note the O-ring

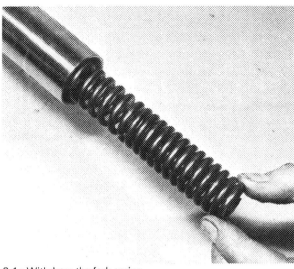

3.1c Withdraw the fork spring

3.2a Using a simple device such as this ...

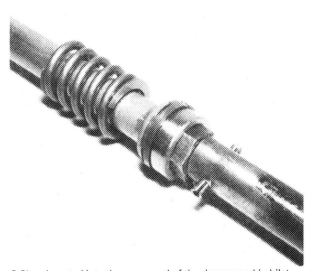

3.2b ... inserted into the upper end of the damper rod (whilst still in the fork leg), the ...

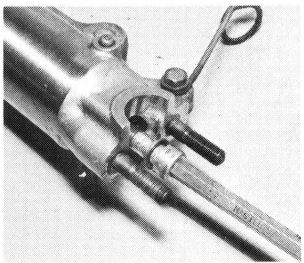

3.2c ... socket bolt can be unscrewed from the lower leg

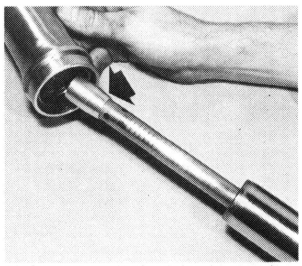

3.2d Separate the stanchion from the lower leg, remove the damper rod seat (arrowed) and ...

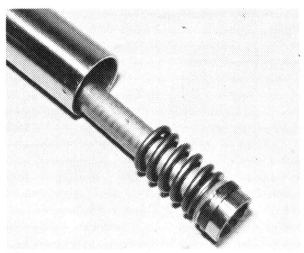

3.2e ... remove the damper rod

3.3a Removal of circlip and a washer (GN model) enables ...

3.3b ... the oil seal to be prised from position

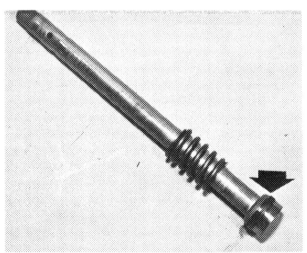

4.5 Check damper rod piston ring (arrowed) and oilways

the housing in the lug or the housing may be damaged. Before reassembling the fork yoke sub-assembly, grease the steering head bearings thoroughly with a multi-purpose grease. There is no provision for further lubrication after assembly.

6 Front forks: replacement

1   Replace the front forks by following in reverse the dismantling procedures described in Sections 2 and 3 of this Chapter. When replacing the fork springs, note that they are of variable pitch. The more closely wound coils should be towards the **top** of the fork leg. Before fully tightening the front wheel spindle clamp bolts (single clamping pinch bolt on the GLT model) and the fork yoke pinch bolts, bounce the forks several times to ensure they work correctly and settle down into their original positions. Complete the final tightening from the wheel spindle upwards.

2   Check that the drain plugs have been re-inserted and tightened. Before replacing either the valve units or the spring adjuster piece, depending on which type of fork is being rebuilt, refill each fork leg with the correct quantity and specification of fork oil. Suzuki recommend that the damping fluid be of SAE 10W/20 viscosity or alternatively, fork oil.

3   If the fork stanchions prove difficult to relocate through the fork yokes, make sure their outer surfaces are clean and

polished, and additionally, use a liberal coating of washing-up liquid, to ease their relocation. This will allow the stanchions to slide more easily. It can also often be advantageous to use a screwdriver blade to open up the clamps, as the stanchions are moved upwards into position. Care should be taken, however, not to apply too much force or the casting may crack.

4   Before the machine is used on the road, check the adjustment of the steering head bearings. If they are too slack, judder will occur especially during braking. There should be no detectable play in the head races when the handlebars are pulled and pushed with the front brake applied hard.

5   Overtight head races are equally undesirable. It is possible to unwittingly apply a loading of several tons on the head races when they have been overtightened, even though the handlebars appear to turn quite freely. Overtight bearings will make the machine roll at low speeds and give generally imprecise handling with a tendency to weave. Adjustment is correct if there is no perceptible play in the bearings and the handlebars will swing to full lock in either direction, when the machine is on the centre stand with the front wheel clear of the ground. Only a slight tap should cause the handlebars to swing. After adjustment has been made by using the peg nut on the steering stem, the upper centre nut should be tightened to the recommended torque setting of 3.6 – 5.2 kgf m (26.0 – 37.5 lbf ft), followed by the pinch bolt to the rear of the upper yoke, to a setting of 1.5 – 2.5 kgf m (11.0 – 18.0 lbf ft).

6   Similarly, before attempting to ride a machine equipped

with the air-assisted forks, the fork legs must be pressurized with the required amount of air. With the valves installed in the top of each fork leg, proceed to pump up to the required pressure. Suzuki recommend the use of a bicycle-type pump, with a suitable adaptor to accept the Shraeder valve, in conjunction with their air pressure gauge (Suzuki part No. 09940-44110). An alternative method is to use the same bicycle-type pump with the simple pressure gauge (Suzuki Part No. 96200-41330) supplied with these machines. A garage air line should only be used in an emergency, as it is not possible to meter the amount of air entering the forks with enough control to obtain the minimal quantity required within the forks. There is also a good chance that the fork oil seals will fail due to the sudden excessive pressure exerted upon them. The front wheel must be clear of the ground when carrying out this process.

7   Note that only air or nitrogen should be used to pressurize the forks. Never attempt to use oxygen or any gas which explodes under pressure. Suzuki recommend a pressure of 0.6 kg/cm^2 (8.5 psi) as a base figure from which to adjust the level of air-assistance, according to the operators' personal preference. The pressure may be allowed to rise to 2.5 kg/cm^2 (35 psi) safely, for a short space of time. The normal working pressure range, however, is between 0.6 and 1.2 kg/cm^2 (8.5 and 17 psi). Exceeding the maximum specified pressure loading will damage fork seals and possibly promote actual fork damage. Always ensure that the difference between the two fork leg pressures is minimal, and never more than 0.1 kg/cm^2 (1.4 psi).

8   When adjusting the front fork air pressure, it must be borne in mind that the rear suspension must also be adjusted to preserve the balance between front and rear damping. It is possible, through total mismatching of the operation of front and rear suspensions, to upset the normal geometry of the frame, to such an extent as to render the machine highly unstable. See Section 13 of this Chapter for rear suspension adjustment procedure.

7 Steering head lock

1   The steering head lock on the GS 850 models is incorporated in the ignition switch mounted on the upper fork yoke between the base of the speedometer and tachometer. The lock, when in a locked position, has a tongue which extends from the body of the lock, when the handlebars are turned to the left or right, and abuts against a plate welded to the base of the steering head. The lock is operated when the ignition switch is turned to the lock or park position.

2   If the lock malfunctions it must be renewed; repair is impracticable. When the lock is changed the key must be changed too to match the new lock.

8 Frame: examination and renovation

1   The frame is unlikely to require attention unless accident damage has occurred. In some cases, replacement of the frame is the only satisfactory course of action if it is badly out of alignment. Only a few frame repair specialists have the jigs and mandrels necessary for resetting the frame to the required standard of accuracy and even then there is no easy means of assessing to what extent the frame may have been overstressed.

2   After the machine has covered a considerable mileage, it is advisable to examine the frame closely for signs of cracking or splitting at the welded joints. Rust can also cause weakness at these joints. Minor damage can be repaired by welding or brazing, depending on the extent and nature of the damage.

3   Remember that a frame which is out of alignment will cause handling problems and may even promote speed wobbles. If misalignment is suspected, as the result of an accident, it will be necessary to strip the machine completely so that the frame can be checked and if necessary, renewed.

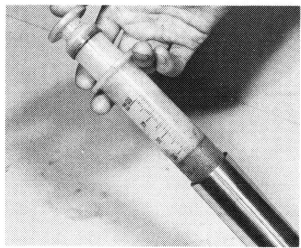

6.2 Refill each leg with correct quantity of oil

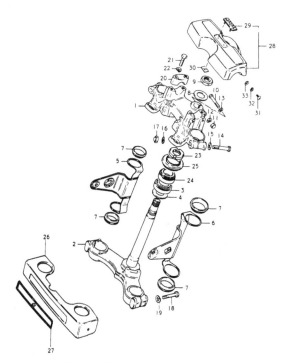

Fig. 4.3 Steering head assembly

1	Upper (crown) yoke	17	Nut – 2 off
2	Lower yoke/steering stem	18	Bolt – 4 off
3	Lower bearing	19	Spring washer – 4 off
4	Washer	20	Handlebar clamp – 2 off
5	Right-hand headlamp bracket	21	Bolt – 4 off
6	Left-hand headlamp bracket	22	Spring washer – 4 off
7	Cushion – 4 off	23	Bearing adjuster ring
8	Washer	24	Upper bearing
9	Nut	25	Dust seal
10	Pinch bolt	26	Lower yoke cover
11	Nut	27	Emblem
12	Spring washer	28	Upper yoke cover
13	Washer	29	Emblem
14	Pinch bolt – 2 off	30	Nut – 2 off
15	Washer – 2 off	31	Screw – 4 off
16	Spring washer – 2 off	32	Spring washer – 4 off
		33	Washer – 4 off

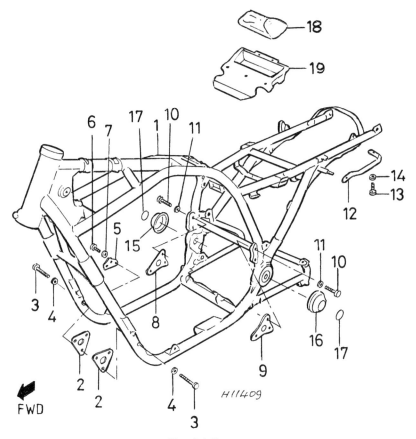

Fig. 4.4 Frame

1	Frame	6	Bolt – 2 off	10	Bolt – 4 off	15	Right-hand frame cover
2	Engine front mounting plate – 2 off	7	Spring washer – 2 off	11	Spring washer – 4 off	16	Left-hand frame cover
3	Bolt – 4 off	8	Right-hand rear engine plate	12	Grab rail	17	Insert – 2 off
4	Spring washer – 4 off	9	Left-hand rear engine plate	13	Bolt	18	Tool roll
5	Engine mounting plate			14	Washer	19	Tool tray

9 Swinging arm bearings: checking and adjustment

1   The rear swinging arm fork pivots on two taper roller bearings, which are supported on adjustable screw stubs fitted to the lugs either side of the frame. After a period of time the taper roller bearings will wear slightly, allowing a small amount of lateral shake at the rear wheel. This condition will lead to a noticeable effect on handling unless adjustment is made.

2   To check the play accurately, and if necessary to make suitable adjustments, it will be necessary to remove the rear wheel and detach the rear suspension units.

3   Place the machine on the centre stand so that the rear wheel is clear of the ground. Remove the rear wheel by following the procedure described in Chapter 5, Section 10.

4   Detach the two rear suspension units by removing their upper mounting stud nuts, and lower mounting bolts.

5   Suspend the caliper unit from the right-hand rear suspension unit upper mounting stud. If care is taken during dismantling, disconnection of the rear brake hydraulic pipe will not be required. This will facilitate reassembly. Carefully ease the hydraulic pipe and the grommet through which it passes from position in the locating clip welded to the swinging arm.

6   The swinging arm fork can now be checked for play. Grasp the fork at the rear end and push and pull firmly in a lateral direction. Any play will be magnified by the leverage effect. Move the swinging arm up and down as far as possible. Any roughness or tightness at one point may indicate bearing damage. If this is suspected, the bearings should be inspected after removal of the swinging arm as described in the following Section.

7   If play is evident, remove the plastic cover plugs from the adjuster studs either side of the frame and slacken off the adjuster locknuts. Using a vernier gauge, check that the distances between each end of the swinging arm cross-member and the adjacent part of the frame are exactly the same. If there is a difference in the width of these gaps, adjustment must be made. Slacken the adjuster stub $\frac{1}{2}$ of a turn on the side with the larger gap, and then tighten the opposite stub to 0.35 – 0.45 kgf m (2.5 – 3.0 lbf ft). Check the gap again and, if necessary, readjust using the same procedure. When using a torque wrench the load on the bearings is adjusted automatically so that all the play is taken up. Where no torque wrench is available, centralise the swinging arm and then tighten the stubs an equal amount, about $\frac{1}{8}$ of a turn at a time. Whilst tightening, move the swinging arm up and down until it can be felt that the bearings are beginning to drag. It is at this point that adjustment is correct.

8   After adjustment is completed, tighten the locknuts without allowing the stubs to rotate. The locknuts should be tightened to the recommended torque setting of 11.0 – 13.0 kgf m (79.5 – 94.0 lbf ft). Refit the plastic plugs.

9   Reassemble the rear wheel and brake and suspension components by reversing the dismantling procedure.

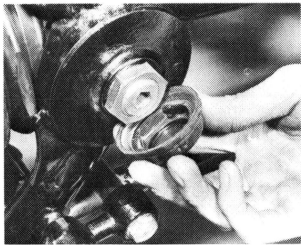
9.7a Remove the cover plugs from each side of the frame

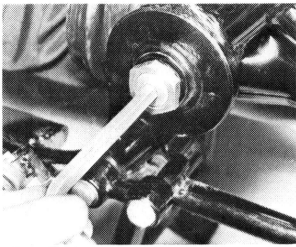
9.7b Use Allen key to adjust swinging arm play

9.8 Tighten the locknuts after adjustment has been made

Exercise care to prevent the inner races of the taper roller bearings, at each side of the swinging arm front cross-member, from dropping free.

5   Take out the bearing inner races and clean and inspect them thoroughly. Clean the outer races whilst they are still in place. Check the rollers for pitting and the outer races for pitting and indentations, and look for a smooth, polished appearance. If the bearings need replacing, the inner races may be simply levered from position. If the bearing outer races require renewal, they must be forcibly extracted from position. Suzuki recommend the use of special tool No. 09941-64510 to remove the outer races, and a further special tool No. 09924-74510, in order to refit new components successfully. It is for this reason that the aid of the local Suzuki Service Agent should be enlisted to carry out this operation.

6   When reassembling, clean and lubricate the bearings thoroughly using a waterproof grease of the type recommended for wheel bearings. No provision is made for further lubrication after assembly.

7   Refit the swinging arm by reversing the dismantling procedure. Adjust the bearings by referring to Section 9, paragraphs 6 and 7.

10 Swinging arm: renewal, renovation and replacement

1   If on inspection for play in the swinging arm bearings it is found that damage has occurred, or wear is excessive, the swinging arm fork should be removed. Commence by following paragraphs 3 and 4 of the preceding Section and then detach the swinging arm as follows.

2   Release the large screw clip securing the rubber gaiter to the boss to the rear of the engine. Prise back the gaiter, so that access may be gained to the final driveshaft flange. Slacken evenly and remove the four flange bolts, turning the flange as necessary to reach the bolts.

3   The final drive gear casing may be detached from the rear wheel as a complete unit after removing the three flange nuts and washers from the mounting studs. Support the considerable weight of the casing as the nuts are removed. Drainage of the lubricating oil is not required, provided the case is removed and stored in an upright position.

4   Loosen the locknuts on the swinging arm adjuster stubs after detaching the plastic covers from either side of the frame. Unscrew the stubs completely and lift the swinging arm fork, complete with final driveshaft, away to the rear of the machine.

10.3a Remove the three flange nuts to enable ...

10.3b ... separation of the final drive case and left-hand side swinging arm member

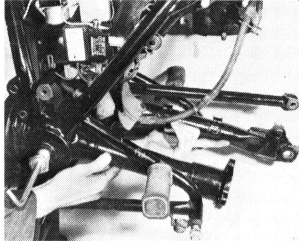

10.4 Lift swinging arm fork out rearwards

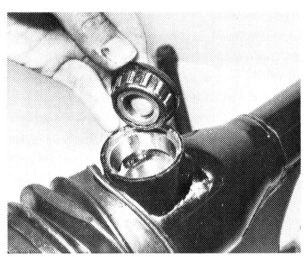

10.5a Clean and inspect inner bearing races

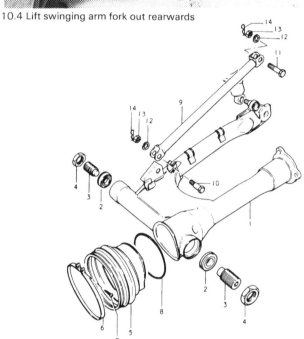

Fig. 4.5 Swinging arm fork

1   Swinging arm
2   Bearing – 2 off
3   Pivot shaft/adjuster –
    2 off
4   Nut – 2 off
5   Dust cover
6   Screw clip
7   Screw
8   Spring clip
9   Torque arm
10 Bolt
11 Bolt
12 Spring washer – 2 off
13 Nut – 2 off
14 R-pin – 2 off

10.5b Check the outer races in situ

11 Final driveshaft: examination and renewal

1   In due course the joint on the upper end of the driveshaft will wear. The joint is of the constant velocity type, sealed for life with its own lubrication reservoir, and hence should have an extremely long life expectancy. The final driveshaft and joint may be removed from the swinging arm unit, after the latter component has been removed from the machine; see Section 10, paragraphs 1 to 4.

2   Using two screwdrivers as levers, with the lever ends placed behind the splined lower boss, withdraw the shaft from the swinging arm. The CV joint may be lifted from the forward end. Wear of the joint will be self-evident, giving a rough,

notchy feel when it is flexed.

3   Check the condition of the splines at both ends of the shaft. If damage or excessive wear is evident, the shaft should be renewed. Inspect the condition of the oil seal and renew it, if required.

4   When refitting the shaft, lubricate the splined ends with a graphite grease.

12 Final drive bevel gear: examination and renovation

1   In much the same way as are the secondary drive and driven bevel gear units, the final drive bevel gear unit is beyond the scope of this manual, and the majority of amateur mechanics. Wear or damage may be indicated by a high pitched whine. Backlash between the crownwheel and pinion may be assessed by holding the output splined boss and rotating the input shaft, first one way and then the other.

2   Failure of the seals either at the input shaft or output shaft will be self-evident by oil leakage.

3   Check the splines on the input shaft and output boss for wear or damage.

4   If attention to the final drive bevel gear unit is required, the complete unit should be returned to a reputable Suzuki Service Agent who should have the necessary special tools and experience to carry out inspection, adjustments and overhaul.

11.1 Examine the constant velocity upper end joint

11.2a Withdraw the final drive shaft for examination

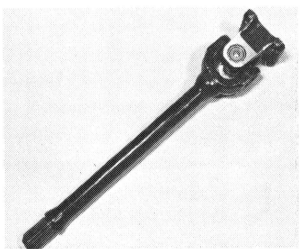

11.2b The driveshaft assembly will give good length of service

12.3a Check condition of splines in drive gear case input and ...

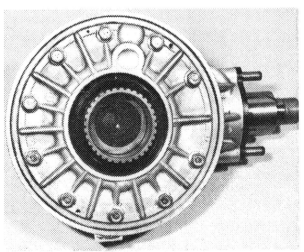

12.3b ... output shafts. Grease all splines before reassembly

13 Rear suspension units: examination and adjustment

1   Rear suspension units of the hydraulically damped types are fitted to the Suzuki GS 850 models. They can be adjusted to give five different spring loadings, without removal from the machine. In addition to this adjustment, the rear suspension incorporates a 4-way damping adjustment facility.

2   Each rear suspension unit has two peg holes immediately above the adjusting notches, to facilitate spring tension adjustment. Either a C-spanner or the screwdriver supplied with the original tool kit can be used to turn the adjusters. Turn clockwise to increase the spring and stiffen up the suspension springing. The recommended settings are:

Position 1 (least tension) for normal solo riding and
Position 5 (greatest tension) for high speed riding and/or when carrying a heavy load.

The intermediate settings, Positions 2 – 4, offer additional settings for varying conditions, as required.

3   The adjustment of the damping force of the rear suspension is carried out in a similar fashion to the spring tension adjustment. A rubber collar is fitted to the top of each suspension unit. Prise this collar free from its normal position to reveal the slotted damper adjustment wheel. Insert the blade of a screwdriver into one of the adjuster slots and turn clockwise to increase the amount of damping force. The positions are numbered from 1 to 4, by stamp marks on the adjuster wheel. As the adjuster wheel is turned, there is a positive click stop as each position is engaged; ensure that the wheel is in the required detent position because the damping effect on any in between setting corresponds to number 4 (stiffest). The softest setting is number 1. On US models both spring and damper standard settings are number 1, while for UK models the standard spring setting is position 2 and the standard damper setting is number 3. The amount of damping force must always correspond accurately with the spring tension selected. See the accompanying table. This enables correct use of the machine, both at high speed and with a passenger and/or heavy load added, whilst maintaining stability and a certain amount of comfort. Care must also be taken to ensure that the damping force rate, and spring tension settings are exactly the same on both rear suspension units, as again, instability will result from a failure to do this.

4   Having adjusted the rear suspension to the required settings for the general usage of the machine, the front suspension air-assistance rate, on the US market models, must now be checked. Re-adjusting either front or rear suspension alone is not recommended as the handling properties of the machine will again be impaired, possibly to a dangerous degree due to a serious imbalance occurring between front and rear suspension operations. See the accompanying table for the recommended front fork air pressure and rear suspension spring and damper settings.

5   The suspension units are sealed and there is no means of topping up or changing the damping fluid. If the damping fails or if the unit leaks, renewal is necessary. If renewal is the only course available, the units must be treated, and replaced, as a matched pair.

14 Centre stand: examination

1   The centre stand is retained on the underside of the frame by two bolts which serve as pivot shafts. A bush is fitted to each shaft. The pivot assemblies on centre stands are often neglected with regard to lubrication and this will eventually lead to wear. It is prudent to remove the pivot bushes from time to time and grease them thoroughly. This will prolong the effective life of the stand.

2   Check that the return spring is in good condition. A broken or weak spring may cause the stand to fall whilst the machine is being ridden, and catch in some obstacle, unseating the rider.

13.3 Prise up rubber collar at top of rear suspension unit and use screwdriver blade to facilitate damping adjustment

Spring setting	Damper setting
I	1 or 2
II	2 or 3
III	3 or 4
IV	3 or 4
V	4

Fig. 4.6 Rear suspension spring and damper setting table

Spring setting	Damper	Front fork air pressure
I	1	$0.6-0.8$ kg/cm^2 (8.5$-$11 psi)
I	2	$0.8-0.9$ kg/cm^2 (11$-$13 psi)
II	2	$0.8-0.9$ kg/cm^2 (11$-$13 psi)
II	3	$0.8-0.9$ kg/cm^2 (11$-$13 psi)
III	3	$1.0-1.1$ kg/cm^2 (14$-$16 psi)
III	4	$1.0-1.1$ kg/cm^2 (14$-$16 psi)
IV	3	$1.0-1.1$ kg/cm^2 (14$-$16 psi)
IV	4	$1.0-1.1$ kg/cm^2 (14$-$16 psi)
V	4	1.2 kg/cm^2 (17 psi)

Rear shock absorber		Front fork spring
Spring setting	Damper	Setting
II	3	I
II	3	II
III	3	II
III	3	III
IV	3	III

Fig. 4.7 Front and rear suspension setting tables

15 Side stand: examination

1   The side stand bolts to a mounting plate attached to the rear of the left-hand lower frame tube. An extension spring ensures that the stand may be retracted when the weight of the machine is taken off the stand.

2   Check that the pivot bolt is secure and that the extension spring is in good condition and not over-stretched. An accident is almost inevitable if the stand extends whilst the machine is on the move.

16 Footrests: examination and renovation

1   Each footrest is an individual unit. The two front footrests are bolted to the frame lower section by two bolts on each side. The two pillion footrests are each retained by a single bolt to a triangulated section rear sub-frame.

2   Both pairs of footrests are pivoted on clevis pins and spring loaded in the down position. If an accident occurs, it is probable that the footrest peg will move against the spring loading and remain undamaged. A bent peg may be detached from the mounting, after removing the clevis pin securing split pin and the clevis pin itself. The damaged peg can be straightened in a vice, using a blowlamp flame to apply heat to the area where the bend occurs. The footrest rubber will, of course, have to be removed as the heat will render it unfit for service. Depending upon the severity of the impact during the machine's horizontal excursion down the road, the footrest rubber will probably be in need of renewal in any case.

17 Brake pedal: examination

1   The rear brake pedal is secured by a single pinch bolt to the splined brake pivot shaft. In the event of damage, the pedal may be removed and treated similarly to a bent footrest, as described in the previous section.

2   The pedal may be pulled off the splined shaft after removing completely the pinch bolt. The outer end of the brake return spring, which shares the pivot shaft, should be displaced from the anchor peg on the frame, so that it may be removed at the same time as the pedal.

18 Dualseat: removal and replacement

1   The dualseat is attached to two lugs on the left-hand side of the frame by means of two pivots (not GLT model) and secured by two clevis pins and split pins. On the GLT model, the seat is not pivoted, but has two lugs on the underside forward part of the seat, which locate with retaining hook on each frame side rail.

2   If it is necessary to remove the dualseat, withdraw the two split pins, take out the clevis pins, and the seat will lift off as a complete unit. On the GLT model, seat removal is an even simpler matter. Insert the ignition key in the lock at the rear of the seat hump, turn clockwise, and lift away the seat by raising it up at the rear, and then unhooking it from the forward retaining hook tabs.

19 Speedometer and tachometer heads: removal and replacement

1   The speedometer and tachometer are mounted together on a single panel on top of the front forks. Each instrument does, however, have a separate holder which secures the instruments to the shared base plate. The separate holders are held to the instruments by two studs projecting from the base of each instrument, and then passing through the shared base plate and into the base cover. A separate plastic case is fitted around the two instruments. The warning light console situated between the speedometer and tachometer, and the gear indicator unit situated between the base of the two instruments, can both be detached once the plastic instrument base has been removed. This case can be detached separately from the rest of the instruments, if required, once the trip-meter reset knob has been removed from the side of the speedometer unit. The warning lamp console is retained by two screws, as is the gear position indicator unit.

2   The instruments may be detached from the machine as a unit, after disconnecting the drive cables and fuel gauge and warning light etc leads at the block connectors. After removing the base plate, the bulb holders may be pulled from position. On the GLT model, the mounting of the instruments is similar to that on the other models. Each instrument does, however, have a separate case instead of the one-piece unit used on the other models.

3   If either instrument fails to record, check the drive cable first before suspecting the head. If the instrument gives a jerky response it is probably due to a dry cable, or one that is trapped or kinked.

4   The speedometer and tachometer heads cannot be repaired by a private owner, and if a defect occurs a new instrument has to be fitted. Remember that a speedometer in correct working order is required by law on a machine in the UK and also in many other countries.

5   Speedometer and tachometer cables are only supplied as a complete assembly. Make sure the cables are routed correctly through the clamps provided on the top fork yoke, brake branch pipe, frame, and the left-hand, lower fork leg guide.

19.1 Remove two domed nuts on each instrument to allow base plate to be detached

20 Speedometer and tachometer drive cables: examination and maintenance

1   It is advisable to detach both cables from time to time in order to check whether they are lubricated adequately, and whether the outer coverings are compressed or damaged at any point along their run. Jerky or sluggish movements can often be attributed to a cable fault.

2   For greasing, withdraw the inner cable. After wiping off the old grease, clean with a petrol-soaked rag and examine the cable for broken strands or other damage.

3   Regrease the cable with high melting point grease, taking care not to grease the last six inches at the point where the cable enters the instrument head. If this precaution is not observed, grease will work into the head and immobilise the movement.

4   If either instrument ceases to function, suspect a broken cable. Fit a complete new cable to restore operating function.

21 Speedometer and tachometer drives: location and examination

1 The speedometer is driven from a gearbox fitted to the front wheel spindle on the left-hand side of the hub. Drive is transmitted through a dog plate fixed to the hub which engages with the drive gear in the gearbox.

2 Provided that the gearbox is repacked with grease from time to time very little wear should be experienced. In the event of failure, the complete gearbox should be renewed. The tachometer drive is taken from the cylinder head cover, between number three and four cylinders. The drive is taken from the overhead camshaft by means of skew-cut pinions, and then by a flexible cable to the tachometer head. It is unlikely that the drive will give trouble during the normal service life of the machine, especially since it is fully enclosed and effectively lubricated.

22 Cleaning the machine

1 After removing all surface dirt with rag or sponge which is washed frequently in clean water, the machine should be allowed to dry thoroughly. Application of car polish or wax to the cycle parts will give a good finish, particularly if the machine receives this attention at regular intervals.

2 The plated parts should require only a wipe with a damp clean rag, but if they are badly corroded, as may occur during the winter when the roads are salted, it is permissible to use one of the proprietary chrome cleaners. These often have an oily base which will help to prevent corrosion from recurring.

3 If the engine parts are particularly oily, use a cleaning compound such as Gunk or Jizer. Apply the compound whilst the parts are dry and work it in with a brush so that it has an opportunity to penetrate and soak into the film of oil and grease. Finish off by washing or hosing down liberally, taking care that water does not enter the carburettors, air cleaner, or the electrics.

4 If possible, the machine should be wiped down immediately after it has been used in the wet, so that it is not garaged under damp conditions which will promote rusting. Remember there is less chance of water entering the control cables and causing stiffness if they are lubricated regularly as described in the Routine Maintenance Section.

23 Fault diagnosis: frame and forks

Symptom	Cause	Remedy
Machine is unduly sensitive to road conditions	Forks and/or rear suspension units have defective damping	Check oil level/air pressure (where applicable) in front forks. Check rear suspension units for incorrect damping/spring settings. Adjust as necessary.
Machine tends to roll at low speeds	Steering head bearings overtight or damaged	Slacken bearing adjustment. If no improvement, dismantle and inspect bearings.
Machine tends to wander, steering is imprecise	Worn or maladjusted swinging arm bearings	Check and if necessary renew bearings. Check adjustment prior to renewal.
Fork action stiff	Fork legs have twisted in yokes or have been drawn together at lower ends	Slacken off spindle nut clamps, pinch bolts in fork yokes and fork top nuts. Pump forks several times before retightening from bottom.
Forks judder when front brake is applied	Worn fork legs and stanchions. Steering head bearings too slack	Renew one or both items. Readjust to take up play.
Wheels out of alignment	Frame distorted as result of accident damage	Check frame alignment after stripping out. If bent, specialist repair is necessary.

Chapter 5 Wheels, brakes and tyres

Refer to Chapter 7 for information relating to the 1981 to 1988 models

Contents

Specifications

Tyres
Front .... 3.50H x 19 – 4PR
Rear .... 4.50H x 17 – 4PR

Tyre pressures

	Solo	Pillion
Front	25 psi (1.75 kg/cm^2)	25 psi (1.75 kg/cm^2)
Rear	28 psi (2.00 kg/cm^2)	32 psi (2.25 kg/cm^2)

For continuous high-speed riding, the pressures should be increased to:

	Solo	Pillion
Front	28 psi (2.00 kg/cm^2)	28 psi (2.00 kg/cm^2)
Rear	32 psi (2.25 kg/cm^2)	40 psi (2.80 kg/cm^2)

Note: pressures apply to original equipment tyres – if non-standard tyres are fitted check with tyre manufacturer or supplier whether different pressures are necessary

Minimum recommended tread depth
Front .... 1.6 mm (0.06 in)
Rear .... 2.00 mm (0.08 in)

Wheel rim runout (maximum)
Front and rear .... 2.0 mm (0.08 in) Axial and radial

Brakes
Front .... Twin hydraulically operated disc
Rear .... Single hydraulically operated disc

Disc thickness
GS 850 GN model:
    Front .... 5.9 – 6.1 mm (0.23 – 0.24 in)
    Service limit .... 5.5 mm (0.22 in)
    Rear .... 6.5 – 6.9 mm (0.26 – 0.27 in)
    Service limit .... 6.0 mm (0.24)
GS 850 GT/GLT models:
    Front .... 5.0 ± 0.2 mm (0.20 ± 0.008 in)
    Service limit .... 4.5 mm (0.18 in)
    Rear .... 6.7 ± 0.2 mm (0.26 ± 0.008 in)
    Service limit .... 6.0 mm (0.24 in)

Disc run out (maximum)

Front and rear .... 0.30 mm (0.012 in)

Brake fluid specifications .... DOT 3 or 4 (USA), SAE J1703 (UK)

Main torque wrench settings

Front spindle:

Clamp nuts .... 1.5 – 2.5 kgf m (11.0 – 18.0 lbf ft)

Spindle nut .... 3.6 – 5.2 kgf m (26.0 – 37.5 lbf ft)

Front caliper mounting bolts .... 2.5 – 4.0 kgf m (18.0 – 29.0 lbf ft)

Rear spindle nut .... 8.5 – 11.5 kgf m (61.5 – 83.0 lbf ft)

Rear caliper mounting bolts .... 2.0 – 3.0 kgf m (14.5 – 21.5 lbf ft)

1 General description

All models within the range are fitted with a 19 inch diameter wheel at the front and an 17 inch diameter wheel at the rear. The front tyre has a 3.50 inch section, and the rear tyre a 4.50 inch section. All the models are fitted with cast alloy wheels as standard equipment. The original equipment tyres are manufactured in Japan, and have an 'H' speed rating, safe up to speeds of 130 mph.

Twin hydraulically-operated disc brakes are fitted at the front, and a single disc brake on the rear wheel.

2 Front wheel: examination and renovation

1   With the cast alloy wheels, a careful check must be made for cracks and chipping, particularly at the spoke bases and the edge of the rim. As a general rule, a damaged cast wheel must be renewed, as cracks will cause stress points which may lead to sudden failure under heavy load. Small nicks may be radiused carefully with a fine file and emery paper (No. 600 – No. 1000) to relieve the stress. If there is any doubt as to the condition of the wheel, advice should be sought from a Suzuki repair specialist.

2   Each wheel is covered with a coating of lacquer, to prevent corrosion. If damage occurs to the wheel, and the lacquer finish is penetrated, the bare aluminium alloy will soon start to corrode. A whitish grey oxide will form over the damaged area, which in itself is a protective coating. This deposit, however, should be removed carefully, whenever possible, and a new protective coating of lacquer applied.

3   Check the lateral run-out at the rim by spinning the wheel and placing a fixed pointer close to the rim edge. If the maximum run-out is greater than 2.0 mm (0.08 mm), Suzuki recommend that the wheel be renewed. This is, however, probably slightly over-cautious; a run-out somewhat greater than this can probably be accommodated without noticeable effect on steering. No means is available for straightening a warped wheel without resorting to the expense of having the wheel skimmed on all faces. If warpage has occurred as the result of impact during an accident, the safest measure is to renew the wheel. Worn wheel bearings may cause rim run-out. These should be renewed as described in Section 8 of this Chapter.

3 Front wheel: removal and replacement

1   Place the machine on the centre stand so that it is resting securely on firm ground with the front wheel well clear of the ground. If necessary, place wooden blocks, below the crankcase to raise the wheel. An alternative is to place a jack below the crankcase to take most of the machine's weight when the front wheel is removed.

2   With the twin front brake calipers utilised on all the GS 850 models, one of the calipers must be detached from the fork leg to allow clearance for the wide section tyre. Remove the caliper as a complete unit – without disconnecting the hydraulic fluid hose – by unscrewing the two bolts which pass through the caliper support bracket and fork leg.

3   Disconnect the speedometer at the gearbox by unscrewing the knurled ring. Pull the cable through the guide clip. Displace the split pin from the wheel spindle nut and slacken the nut slightly. On the GN and GT models the wheel may be removed either by detaching the two spindle clamps or by slackening the clamp bolts and withdrawing the spindle. In the latter case the speedometer gearbox and wheel spacer will fall free as the wheel is lowered from place. On the GLT model, the single spindle pinch bolt at the bottom of the right-hand fork leg, should be slackened, prior to removal of the split pin and spindle nut. The spindle can then be removed to the left-hand side of the machine.

4   When refitting the wheel into the forks, ensure that the speedometer gearbox is fitted with the embossed arrow-mark pointing upwards. Also ensure that the wheel spacer is refitted correctly; the flat face fits against the bearing. Tighten the wheel spindle nut fully to a torque setting of 3.6 – 5.2 kgf m (26.0 – 37.5 lbf ft) before tightening either the single pinch bolt (GLT) or the four spindle clamp nuts. The recommended torque setting for these nuts is 1.5 – 2.5 kgf m (11.0 – 18.0 lbf ft). Do not omit the split pin, replacing it with a new item, or an R-pin, if it is at all suspect. The two nuts holding each spindle clamp should be tightened down evenly so that the gap between the clamp and fork leg is equal either side of the wheel spindle.

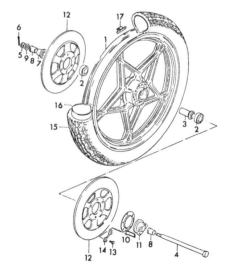

Fig. 5.1 Front wheel – GS 850 GN and GT models

1	Wheel	10	Dust cover
2	Bearing – 2 off	11	Speedometer drive
3	Centre spacer		gearbox
4	Wheel spindle	12	Brake disc – 2 off
5	Castellated nut	13	Bolt – 12 off
6	Split pin	14	Tab washer – 6 off
7	Bearing right-hand spacer	15	Tyre
8	Spacer – 2 off	16	Inner tube
9	Washer	17	Balance weight – A/R

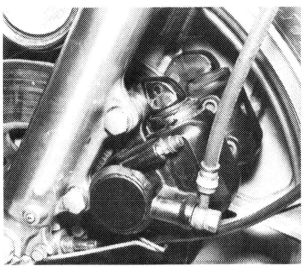

3.2 Remove two large bolts to free caliper

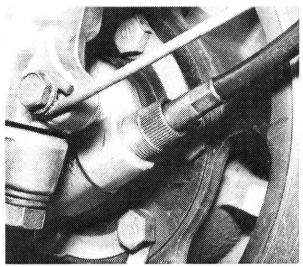

3.3a Disconnect speedometer cable at drive gearbox

3.3b Clamp bolts (GN and GT models) should be slackened to facilitate spindle removal

3.4a Note method of operation of drive gearbox; two tabs locate with two notches and ...

3.4b ... ensure gearbox arrow mark UP is correctly positioned

3.4c Spacer flat face fits against LH bearing

3.4d Do not omit split pin

that was marketed in the UK, and those GN models produced for the US market. In addition both types of GN model caliper, differ significantly from the calipers used on the current GT and GLT models.

6   To gain access to the pads for renewal, the caliper assembly must be detached from the front fork. Removal of the wheel is not required, nor is separation of the caliper from the hydraulic hose. Remove the two bolts which pass through the fork leg into the caliper support bracket and lift the complete caliper unit upwards, off the disc.

7   On the GN models, remove the single screw and either the convolute or flat backing plate, from the inner side of the caliper unit. The inner pad is now free and may be displaced towards the centre of the caliper and lifted out. The outer pad which abuts against the caliper piston is not retained positively and may be lifted out. On the GT and GLT models, the pad replacement operation is even more straightforward. With the caliper separated from the fork leg, remove the brake pad retaining spring from the top of the pads and then pull out both pads and the outer pad shim.

4   Front brake assembly: examination and brake pad renewal

1   Check the front brake master cylinder, hoses and caliper units for signs of leakage. Pay particular attention to the condition of the hoses, which should be renewed without question if there are signs of cracking, splitting or other exterior damage. Check the hydraulic fluid level by referring to the upper and lower level lines visible on the exterior of the transparent reservoir body.

2   Replenish the reservoir after removing the cap on the brake fluid reservoir and lifting out the diaphragm plate. The cap is either a screw-on/off fitting, or is retained by four screws, depending on the model. The condition of the fluid can be checked at the same time. Checking the fluid level is one of the maintenance tasks which should **never be neglected**. If the fluid is below the lower level mark, brake fluid of the correct specification must be added. **Never** use engine oil or any fluid other than that recommended. Other fluids have unsatisfactory characteristics and will rapidly destroy the seals. The fluid level is unlikely to fall other than a small amount, unless leakage has occurred somewhere in the system. If a rapid change of level is noted, a careful check for leaks should be made before the machine is used again. It is also worth noting that Suzuki recommend that the brake hoses should be renewed every two years in the interests of safety.

3   The two sets of brake pads should be inspected for wear. Each has a red groove, which marks the wear limit of the friction material. When this limit is reached, both pads in the set must be renewed, even if only one has reached the wear mark. In normal use both sets of pads will wear at the same rate and therefore both sets must be renewed. To facilitate the checking of brake pad wear on the GT and GLT models, each caliper is provided with an inspection window, closed by a small cover. Prise the cover from position in order to inspect the pads.

4   If the brake action becomes spongy, or if any part of the hydraulic system is dismantled (such as when a hose has been renewed) it is necessary to bleed the system in order to remove all traces of air. Follow the procedure in Section 7 of this Chapter.

5   It is, perhaps, worth noting at this time, that there have been, to date, three different types of calipers fitted to the front forks of GS 850 models. There are, therefore, several small differences between the calipers fitted to the original GN model

4.3 Red line on pad denotes maximum wear limit

4.7a Remove single screw and plate and ...

4.7b ... displace inner pad towards piston pad

4.7c The piston pad may be lifted out

4.8 Do not omit metal shim (where fitted) upon reassembly

8   Refit the new pads and replace the caliper by reversing the dismantling procedure. The caliper piston should be pushed inwards slightly so that there is sufficient clearance between the brake pads to allow the caliper to fit over the disc. It is recommended on the GN models that the outer periphery of the outer (piston) pad is slightly coated with disc brake assembly grease (silicone grease). Use the grease sparingly and ensure that grease **does not** come in contact with the friction surface of the pad. When refitting pads to the GT/GLT models' calipers, the pads should **not** be greased at any point.

9   Ensure that during the pad removal and refitting operation, the brake lever is not operated at any time. This will operate and displace the caliper piston; reassembly is then considerably more difficult.

10 In the interests of safety, always check the function of the brakes; pump the brake lever several times to restore full braking power, before taking the machine on the road.

5 Front brake caliper: dismantling, examination and overhaul

1   Select a suitable receptacle into which may be drained the hydraulic fluid. Remove the banjo bolt holding the hydraulic hose at the caliper and allow the fluid to drain. Repeat this operation with the second caliper unit. Take great care not to allow hydraulic fluid to spill onto paintwork; it is very effective paint stripper. Hydraulic fluid will also damage rubber and plastic components.

2   Remove the caliper from the fork leg and displace the brake pads as described in the preceding Section. With the two calipers fitted, they should be dismantled and reassembled individually, to prevent the accidental interchange of components.

3   Remove the two caliper spindles which pass through the caliper body. These take the form of normal, hexagon-headed bolts on the US market GN model and Allen bolts on the UK variant of the GN model. On the GT/GLT models, the spindles can be withdrawn once the two bolts joining the caliper body to the caliper support bracket have been removed. It will be noted that the spindles on the GN models also serve to retain the caliper to its support bracket. The caliper body should now be separated from its support bracket. Prise out the piston boot, using a small screwdriver, taking care not to scratch the surface of the cylinder bore. The piston can be displaced most easily by applying an air jet to the hydraulic fluid feed orifice. Be prepared to catch the piston as it falls free. Displace the annular piston seal from the cylinder bore groove.

4   Clean the caliper components thoroughly, only in hydraulic brake fluid. **Never** use petrol or cleaning solvent for cleaning hydraulic brake parts otherwise the rubber components will be damaged. Discard all the rubber components as a matter of course. The replacement cost is relatively small and does not warrant re-use of components vital to safety. Check the piston and caliper cylinder bore for scoring, rusting or pitting. If any of these defects are evident it is unlikely that a good fluid seal can be maintained and for this reason the components should be renewed.

5   To assemble the caliper, reverse the removal procedure. When assembling pay attention to the following points. Apply Suzuki caliper grease – a high heat resistance type (Part No. 99000-25100) – to the caliper spindles. Also note the two O-rings fitted on the rearmost of the two spindles. These should be renewed upon reassembly. Apply a generous amount of brake fluid to the inner surface of the cylinder and to the periphery of the piston, then assemble. Do not assemble the piston with it inclined or twisted. When installing the piston push it slowly into the cylinder while taking care not to damage the piston seal. On both types of caliper fitted to the GN models, apply Suzuki brake pad grease around the periphery of the moving pad. On the GT/GLT model calipers, ensure that the caliper spindle rubber boots are not perished or damaged in any other way, and are refitted correctly with the spindles. **Do not** apply

any grease to either of the pads in these calipers. Ensure the small pad guide is refitted, and the anti-squeal shim to the outer face of the outer pad.

6 Ensure that all the retaining bolts throughout the brake assembly are fully tightened. The correct torque setting for the caliper spindle bolts (caliper body to support bracket retaining bolts on the GT/GLT models) is 1.5 – 2.0 kgf m (11.0 – 14.5 lbf ft). The brake hose union bolt should be secured to a torque setting of 1.5 – 2.5 kgf m (11.0 – 18.0 lbf ft). The two caliper

mounting bolts should be tightened down to a recommended torque figure of 2.5 – 4.0 kgf m (18.0 – 29.0 lbf ft).

7 Bleed the brake after refilling the reservoir with new hydraulic brake fluid, then check for leakage while applying the brake lever tightly. After a test run, check the pads and brake discs.

8 Note that any work on the hydraulic system must be undertaken under ultra-clean conditions. Particles of dirt will score the working parts and cause early failure of the system.

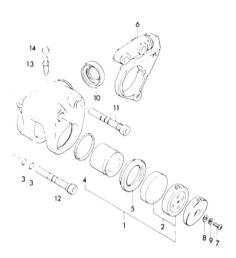

Fig. 5.2 Front brake caliper – GS 850 GN model (UK)

1	Piston/pad set	8	Washer
2	Disc pad set	9	Spring washer
3	O-ring – 4 off	10	Boot
4	Piston seal	11	Spindle
5	Boot	12	Spindle
6	Caliper bracket	13	Bleed nipple
7	Screw	14	Bleed nipple cap

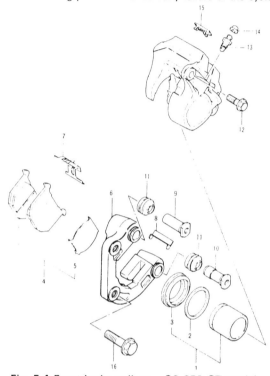

Fig. 5.4 Front brake caliper – GS 850 GT model

1	Piston/seal set	9	Spindle
2	Piston seal	10	Spindle
3	Boot	11	Spindle boot
4	Pad set	12	Bolt – 2 off
5	Shim	13	Bleed nipple – 2 off
6	Caliper bracket	14	Bleed nipple cap – 2 off
7	Pad spring	15	Inspection window
8	Pad guide plate – 2 off	16	Bolt – 2 off

6 Front disc brake master cylinder: examination and renovation

1 The master cylinder and hydraulic reservoir take the form of a combined unit mounted on the right-hand side of the handlebars, to which the front brake lever is attached. The master cylinder is actuated by the front brake lever, and applies hydraulic pressure through the system to operate the front brake when the handlebar lever is manipulated. The master cylinder pressurises the hydraulic fluid in the brake pipe which, being imcompressible, causes the piston to move in the caliper unit and apply the friction pads to the brake. If the master cylinder seals leak, hydraulic pressure will be lost and the braking action rendered much less effective.

2 Before the master cylinder can be removed the system must be drained. Place a clean container below both caliper units and attach a plastic tube from the bleed screw on the top of each caliper unit to the container. Open the bleed screws one complete turn and drain the system by operating the brake lever until the master cylinder reservoir is empty. Close the bleed screw and remove the pipes.

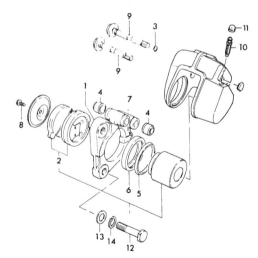

Fig. 5.3 Front brake caliper – GS 850 GN model (US)

1	Piston/pad set – 2 off	8	Screw – 2 off
2	Disc pad set – 2 off	9	Spindle – 4 off
3	O-ring – 2 off	10	Bleed nipple – 2 off
4	Spindle seal – 8 off	11	Bleed nipple cap– 2 off
5	Piston seal – 2 off	12	Bolt – 4 off
6	Boot – 2 off	13	Washer – 4 off
7	Caliper bracket – 2 off	14	Spring washer – 4 off

3   Remove the front brake stop lamp switch from below the master cylinder. Unscrew the union bolt and disconnect the connection between the brake hose and the master cylinder. Unscrew the two master cylinder fastening bolts and remove the master cylinder body from the handlebar. Make sure when separating the brake hose/master cylinder connection, a rag is strategically positioned to absorb any drops of brake fluid. Remove the cap from the reservoir, either simply unscrewing the cap, or, on the GT model only, unscrewing the four retaining screws and lifting the cap. Empty any surplus fluid from the reservoir, again ensuring no fluid finds it way onto paintwork, plastic or rubber components.

4   Remove the brake lever from the body, remove the boot stopper (taking care not to damage the boot) and then remove the boot. Remove the circlip that was hidden by the boot, the piston, primary cup, spring and check valve. Place the parts in a clean container and wash them in new brake fluid. Examine the cylinder bore and piston for scoring. Renew if scored. Check also the brake lever for pivot wear, cracks or fractures and the hose union threads and brake pipe threads for cracks or other signs of deterioration.

5   When assembling the master cylinder follow the removal procedure in reverse order. Pay particular attention to the following points. Make sure the primary cup is fitted the correct way round. Renew the split pin of the brake lever pivot nut and fit it securely. Mount the master cylinder to the handlebar so the gap between the master cylinder and the right-hand handlebar switch cluster is 2 mm (0.08 in) and the reservoir is horizontal when the motorcycle is on the centre stand with the steering in the straight ahead direction. This gap measurement applies to the GN model only. On the GT and GLT models, the required gap is 5 – 7 mm (0.19 – 0.27 in). Note that when fitting the master cylinder, on the GT model only, the upper clamp bolt must be tightened fully before the lower clamp bolt, so that there is no gap at the top of the clamp. A small clearance should be present at the bottom of the clamp; say 2 mm (0.08 in). Fill with fresh fluid and bleed the system. Be sure to check the brake reservoir by removing the reservoir cap. If the level is below the ring mark inside the reservoir, refill to the level with the prescribed brake fluid.

6   The component parts of the master cylinder assembly and the caliper assembly may wear or deteriorate in function over a long period of use. It is however, generally difficult to foresee

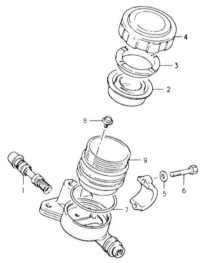

**Fig. 5.5 Front brake master cylinder – GS 850 GN model
(GLT model similar)**

1	Piston/seal assembly	6	Bolt – 2 off
2	Diaphragm	7	O-ring
3	Seal	8	Screw – 2 off
4	Filler cap	9	Reservoir
5	Washer - 2 off		

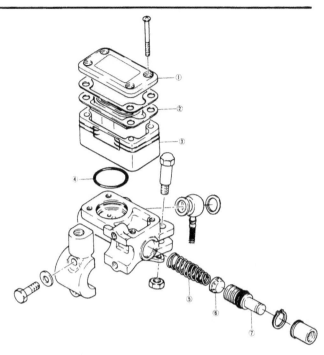

Fig. 5.6 Front brake master cylinder – GS 850 GT model

1	Filler cap	5	Spring
2	Diaphragm	6	Piston cap
3	Reservoir	7	Piston
4	O-ring		

how long each component will work with proper efficiency and from a safety point of view it is best to change all the expendable parts every two years on a machine that has covered a normal mileage.

7   Bleeding the hydraulic brake system

1   If the hydraulic system has to be drained and refilled, if the front brake lever travel becomes excessive or the lever operates with a soft or spongy feeling, the brakes must be bled to expel air from the system. The procedure for bleeding the hydraulic brake is best carried out by two persons.

2   First check the fluid level in the reservoir and top up with fresh fluid.

3   Keep the reservoir at least half full of fluid during the bleeding procedure.

4   Refit the cap on to the reservoir to prevent a spout of fluid or the entry of dust into the system. Commence the bleeding with the left-hand caliper first, followed by the right-hand unit, separately. Place a clean glass jar below the caliper bleed screw and attach a clear plastic pipe from the caliper bleed screw to the container. Place some clean hydraulic fluid in the jar so that the pipe is always immersed below the surface of the fluid.

5   Unscrew the bleed screw one half turn and squeeze the brake lever as far as it will go but do not release it until the bleeder valve is closed again. Repeat the operation a few times until no more air bubbles come from the plastic tube.

6   Keep topping up the reservoir with new fluid. When all the bubbles disappear, close the bleeder valve. Remove the plastic tube and install the bleeder valve dust cap. Check the fluid level in the reservoir, after the bleeding operation has been completed.

7   Reinstall the diaphragm and tighten the reservoir cap securely. Do not use the brake fluid drained from system, since it will contain minute air bubbles.

8   Never use any fluid other than that recommended. Oil must not be used under any circumstances.

8   Front wheel bearings: examination and replacement

1   Access to the front wheel bearings can be made after removal of the speedometer gearbox and spindle spacer.
2   The wheel bearings can be drifted out of position, using a suitable drift. Support the wheel so the exit of the bearing is not obstructed. When the first bearing has been removed the spacer that lies between the two bearings can be removed. Insert the drift and drive out the opposite bearing.
3   Remove all the old grease from bearings and hub. Wash the bearings in petrol and dry them thoroughly. Check the bearings for roughness by spinning them whilst holding the inner track with one hand and rotating the outer track with the other. If there is the slightest sign of roughness renew them.
4   Before driving bearings back into the hub, pack the hub with new grease and also grease the bearings. Use the same double diameter drift to place them in position, ensuring that the bearings enter their position perfectly squarely, otherwise they may twist during fitment making removal very troublesome.
5   Refit any dust covers or oil seals which may have been displaced during the original dismantling operation.

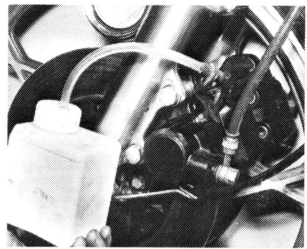
7.4 Clear plastic tube attached to bleed nipple, and container are required to bleed system

8.2a Use a drift to remove bearing, followed by ...

8.2b ... the central internal spacer

9   Removing and replacing the disc

1   It is unlikely that the brake discs will require attention unless bad scoring has developed or the discs have warped. To detach the discs first remove the wheel as described in Section 3 of this Chapter. On GT and GLT models, mark each disc before removal so that it can be correctly refitted; the discs are handed and must not be swapped over. Each disc is retained by six bolts screwed into the hub, which are linked in pairs by tab washers. Bend down the ears of the tab washers and remove the bolts. The discs can then be eased off the hub bosses.
2   The brake disc can be checked for wear and for warpage whilst the front wheel is still in the machine. Using a micrometer, measure the thickness of the disc at the point of greatest wear. If the measurement is much less than the recommended service limit given in the Specifications section of this Chapter, the disc should be renewed. Check the warpage (run out) of the disc by setting up a suitable pointer close to the outer periphery of the disc and spinning the front wheel slowly. If the total warpage is more than 0.30 mm (0.012 in), the disc should be renewed. A warped disc, apart frrom reducing the braking efficiency, is likely to cause juddering during braking and will also cause the brake to bind when it is not in use.
3   Replace the discs by reversing the dismantling procedure.

Ensure that the twelve bolts are tightened fully to the recommended torque setting of 1.5 – 2.5 kgf m (11.0 – 18.0 lbf ft), and that the tab washer ears are bent up against the bolt head flats.

10   Rear wheel: examination, removal and renovation

1   Place the machine on the centre stand so that the rear wheel is raised clear of the ground. Follow the procedures laid down in Section 2 of this Chapter, and check for rim alignment, rim damage etc.
2   To remove the rear wheel of the GS 850 models safely and easily, a specific procedure must be followed. By adopting the procedure that follows, the risks of the machine toppling over, or off its centre stand, are minimised, and damage to the machine can be avoided.
3   Position the machine on firm, level ground, and raise it onto its centre stand. Insert a short length of small diameter metal rod, ie the screwdriver attachment from the original tool kit, into the hole provided to the rear of the right-hand side of the centre stand pivot. This will act as a locking device, and prevent the machine from rolling off the centre stand.
4   Detach both rear suspension units from their upper mounting studs, having removed the domed retaining nuts. This will facilitate vertical movement of the swinging arm/final drive

shaft and gear casing unit.

5   Before progressing further, note that if the machine has been run shortly before the wheel removal operation commenced, the exhaust silencers will be hot to the touch. Exercise care, therefore, to avoid bringing unprotected hands into painful contact with the silencers.

6   The swinging arm fork and gear casing unit must now be raised, in order to bring the rear wheel spindle above the level of the silencers. This may be done by attaching the 14 mm end of one of the open-ended spanners in the original tool-kit to the rearmost caliper mounting bolt, at such an angle that it may be used as a lifting handle. A safer alternative would be to acquire a 14 mm ring end spanner of suitable length; the use of this tool would preclude the chance of the open-end spanner slipping off the bolt head. As the swinging arm fork and gear casing unit is raised up to the required height, insert the steel rod from the original tool-kit through the hollow lug on the pillion footrest sub-frame, and into the support bracket on the swinging arm right-hand member. (See accompanying photograph). This rod will support the swinging arm fork at the height required for easiest removal of the rear wheel spindle.

7   Remove both lower retaining nuts and detach both the rear suspension units from their mounting studs. Remove the split pin and nut securing the torque arm rod to the top of the caliper support bracket. Remove the bolt securing the hydraulic hose guide clamp and free the hose. Remove the split pin or R-clip securing the castellated spindle nut, and slacken and remove the nut. Withdraw the spindle slowly. As the rear caliper and its support bracket assembly becomes free, lift the caliper clear of the rear disc, and suspend the caliper assembly, by the spindle hole in the support bracket, from the upper suspension unit mounting stud. During this time, **do not** allow the hydraulic brake hose to come into contact with the silencer if it is still hot. Withdraw the spindle fully, noting the various spacers which will drop free. Displace the wheel to the right and off the final drive gear casing splines.

8   In order that the wheel and tyre may now be removed from the under-seat area, either the rear of the machine must be raised or the front forks compressed. A third alternative would be to position the machine, at the beginning of the wheel removal sequence, on the edge of a convenient step or kerb so that the wheel can drop free. This method is not recommended by Suzuki, and would really only be possible with the help of an assistant to steady the machine. Suzuki recommend the use of the device supplied with the original tool-kit, to compress the front forks. On US market models, bleed all the air slowly out of the fork legs by depressing the valve cores, with the protector caps removed. Move the handlebars until they are at full right lock, with the forks against the lock stops. Position the fork compression tool so that the main hook is secured below the left-hand front mudguard boss, that is, the lug into which the rearmost, left-hand side mudguard mounting bolt is fitted. Place the upper, loop end, of the tool, over the left-hand side fork stop. Lower the handle of the fork compression tool, until the forks are compressed, and then secure the tool in this position by fitting the second hook around the handle of the tool. With the machine in this attitude, the rear wheel may now be easily rolled clear. It should be borne in mind during this operation, that the procedure just described, although somewhat time consuming, takes considerably less time to actually carry out than it takes to fully describe.

9   Refit the rear wheel by reversing the dismantling procedure, noting certain points. Do not twist or bend the hydraulic brake hose when refitting the caliper and its support bracket, and ensure the rear brake pedal is not operated at any time whilst the caliper is not engaged with the disc. Ensure that all the wheel spacers are refitted correctly. Ensure that the torque arm is secure (see torque settings below) and that the securing nut split pin is replaced. Likewise, do not omit the wheel spindle nut split pin or R-clip, whichever is utilized. Ensure the brake hose retaining clamp is refitted and its securing bolt tightened. Make sure the front forks are recharged with the correct amount of air (US models only). Remove the aids used in supporting and

securing the swinging arm fork and centre stand respectively.

Recommended torque settings:

Rear wheel spindle nut	8.5 – 11.5 kgf m
	(61.5 – 83.0 lbf ft)
Torque arm nut	2.0 – 3.0 kgf m
	(14.5 – 21.5 lbf ft)
Rear suspension unit nuts	2.0 – 3.0 kgf m
	(14.5 – 21.5 lbf ft)

Fig. 5.7 Rear wheel – GS 850 GN model

1	Wheel	14	Washer
2	Damping rubber – 6 off	15	Final drive boss
3	Right-hand bearing	16	Stud bolt – 6 off
4	Left-hand bearing	17	O-ring
5	Bearing spacer	18	Nut – 6 off
6	Caliper spacer	19	Tab washer – 3 off
7	Left-hand spacer	20	Bolt – 6 off
8	Right-hand spacer	21	Brake disc
9	Wheel spindle	22	Bolt – 6 off
10	Castellated nut	23	Tab washer – 3 off
11	Brake caliper bracket	24	Tyre
12	Split pin	25	Inner tube
13	Washer	26	Balance weight – A/R

9.1 Each disc is retained by six bolts secured by locking plates

9.2 Note minimum thickness specification indicated on disc centre

10.3 Use the tool kit screwdriver to lock centre stand as shown

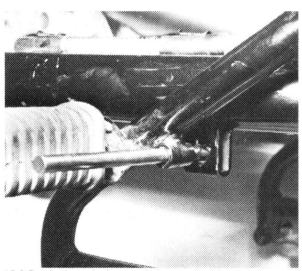

10.6 Retain the swinging arm fork at required height by inserting steel rod as shown

10.7a Remove rear suspension units

10.7b Remove split pin/R-pin and nut, securing torque arm

10.7c Remove bolt securing rear brake hose guide clamp

10.7d With caliper and bracket assembly free, pull it clear of disc and ...

10.7e ... use upper RH suspension unit mounting stud as support for caliper/bracket

10.8 With the front forks compressed (see text), remove wheel

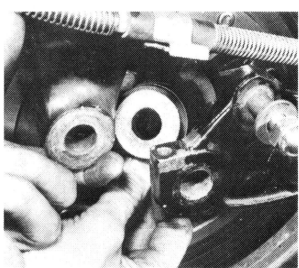

10.9a Ensure the spacers fitted both behind and ...

10.9b ... in front of the caliper bracket are refitted

10.9c Refit brake hose guide clamp bolt and do not omit spindle nut split pin

11 Rear disc brake: examination and pad renewal

1   In general, remarks concerning the front disc brake as described in Section 4 of this Chapter apply equally to the rear disc brake. The rear brake master cylinder/reservoir unit is fitted to the right-hand side of the machine, behind and below the frame cover.

2   To inspect the brake pads for wear, prise off the inspection cap fitted to the top of the caliper. Each pad is stepped slightly, the step nearer the backing plate being painted red. If either pad is worn down to the red line, the pads must be renewed as a set.

3   Pad removal may take place without displacing the caliper unit or the wheel. Pull out the stop pin which passes through each of the two pad mounting pins. Displace one mounting pin and remove the two hair springs. Push out the final pin and lift each pad out individually, removing the outer pad first.

4   Install new pads by reversing the dismantling procedure. If necessary, push back each piston to give the required clearance between the piston and the disc. The shim fitted to the piston side of each pad must be positioned with the punched arrow mark pointing in the direction of wheel travel.

12 Rear disc brake caliper: examination and overhaul

1   Unlike the front brake caliper, which has only one piston, the rear brake caliper has two pistons and two moving brake pads. The general procedure however for dismantling and overhaul is similar to that described for the front brake in Section 5 of this Chapter. Unlike the front calipers, however, where there are three variations in caliper design, the GS 850 models all utilize the same type of rear caliper. The type of pads fitted to the rear caliper do not require grease applying to them at any point.

2   When refitting the caliper to its support bracket, tighten the two retaining bolts to a torque setting of 2.0 – 3.0 kgf m (14.5 – 21.5 lbf ft).

3   After reassembling the caliper, the hydraulic circuit should be bled of all air as described in Section 7. Both sides of the caliper unit should be bled individually by fitting separate bleed pipes to the two nipples provided. Bleed air from the inboard valve first, followed by the outboard unit.

12.1a Pull out the mounting pin stop pins and ...

12.1b ... withdraw the pins to free the pads

12.1c Each pad can then be lifted out

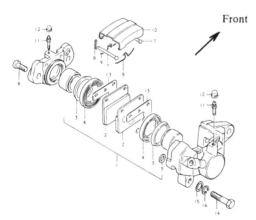

Front

Fig. 5.8 Rear brake caliper – GS 850 GN model

1	Piston/pad set	9	Spring clip – 2 off
2	Pad – 2 off	10	Dust cover
3	Piston seal – 2 off	11	Bleed nipple – 2 off
4	Boot – 2 off	12	Bleed nipple cover – 2 off
5	Seal	13	Shim – 2 off
6	Bolt – 2 off	14	Bolt – 2 off
7	Pad retaining pin – 2 off	15	Washer – 2 off
8	Split pin – 2 off	16	Spring washer – 2 off

Tyre changing sequence - tubed tyres

 Deflate tyre. After pushing tyre beads away from rim flanges push tyre bead into well of rim at point opposite valve. Insert tyre lever adjacent to valve and work bead over edge of rim.

Use two levers to work bead over edge of rim. Note use of rim protectors

 Remove inner tube from tyre

When first bead is clear, remove tyre as shown

 When fitting, partially inflate inner tube and insert in tyre

Work first bead over rim and feed valve through hole in rim. Partially screw on retaining nut to hold valve in place.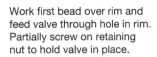

 Check that inner tube is positioned correctly and work second bead over rim using tyre levers. Start at a point opposite valve.

Work final area of bead over rim whilst pushing valve inwards to ensure that inner tube is not trapped

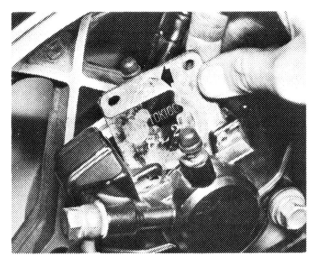

12.1d Do not omit metal pad shim upon reassembly; note direction arrow on shim

13 Rear brake master cylinder: removal and overhaul

1   In common with the unit fitted to the front brake system, the rear brake master cylinder and fluid reservoir are a combined unit. The master cylinder is fitted inboard of the rear frame right-hand triangulation and is operated via a pushrod from the foot brake pedal.

2   To detach the master cylinder, remove the split pin and clevis pin from the lower end of the operating pushrod. Disconnect the hydraulic hose at the master cylinder by removing the banjo bolt, and allow the hose fluid to drain into a suitable container. The master cylinder is secured to the frame by two bolts passing through lugs on the master cylinder body.

3   Remove the reservoir cap and diaphragm and allow the fluid to drain. Pull the gaiter along the pushrod and remove the circlip to allow detachment of the pushrod assembly. Using a wood dowel inserted through the hydraulic fluid inlet orifice push out the piston/seal assembly.

4   Refer to the procedure given in Section 6 for inspection and reassembly details.

5   As an alternative method of removal, should the master cylinder not require overhauling, the complete unit of rear caliper, hydraulic hose, master cylinder and fluid reservoir, can be detached as one assembly. Having detached the master cylinder (see paragraph 2 above) but not disconnected the brake hose, remove the single retaining bolt, and remove the fluid reservoir. Release the brake fluid hose from its retaining clamp on the rear of the right-hand swinging arm member, and then separate the caliper from its support bracket, by removing the two retaining bolts and detaching the torque arm. With some wriggling and manoeuvring of the components around the frame tubes, the complete assembly can be removed.

14 Rear brake pedal height: adjustment

1   The upper limit of travel of the brake pedal may be adjusted by means of the stopper bolt and locknut fitted to the pedal pivot mounting bracket. Slacken the locknut and unscrew slightly the bolt. Loosen the locknut on the operating pushrod, and adjust the pushrod, by moving it upwards or downwards with the headed adjuster provided, until the brake pedal is 20 mm (0.8 in) below the upper edge of the footrest. Secure the pedal in this position by tightening the pushrod locknut fully. Adjust the clearance between the pedal pivot and the return stopper bolt to 0.5 mm (0.02 in) and then tighten the locknut.

2   Adjustment of the pedal may necessitate readjustment of the rear stop lamp switch.

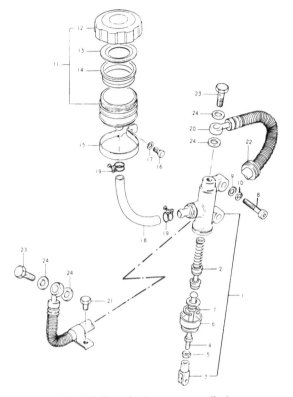

Fig. 5.9 Rear brake master cylinder

1	Master cylinder assembly	13	Seal
2	Piston/seal set	14	Diaphragm
3	Clevis fork	15	Clamp
4	Push rod	16	Bolt
5	Locknut	17	Spring washer
6	Boot	18	Hose
7	Circlip	19	Clip – 2 off
8	Bolt – 2 off	20	Banjo union
9	Washer – 2 off	21	Bolt
10	Spring washer – 2 off	22	Hose grommet – 2 off
11	Reservoir assembly	23	Banjo union bolt – 2 off
12	Cap	24	Sealing washer – 4 off

13.1 Rear brake master cylinder is operated by pushrod and rocking shaft

13.2 Two Allen head bolts (arrowed) secure master cylinder to the frame

13.3 Remove reservoir cap and drain fluid. Note single mounting bolt on frame

15 Rear wheel damper unit: examination and renovation

1   The rear wheel damper unit, or cush drive unit, is contained in the left-hand side of the wheel hub. It takes the form of six rubber bushes fitted over six protruding hexagon-headed stub ends. The headed stubs are fitted through the driven joint on the left-hand side of the hub. The driven joint is in the form of a plate with a splined central bore. The driven joint is retained in position by a series of six bolts, which are secured in pairs by strong tab washers.

2   The function of the damper unit is, as its name suggests, to act as a damping medium when there is a surge or snatch in the transmission at the rear wheel. The result is a smoother delivery of power to the road, due to the stubs on the driven joint being able to move within certain limits against the rubber bushes, and so absorb any shocks being transmitted down the drive shaft. Obviously, therefore, should the rubber bushes become compacted, perished, or start to break up, leading to more movement at the driven joint than is required, the bushes should be renewed.

3   Remove the rear wheel from the machine following the procedure laid down in Section 10 of this Chapter. Place the wheel with the disc side downwards. Straighten the locking washers on the six driven joint retaining bolts, and remove the bolts and washers. Lever up the driven joint to free the six stubs from the rubber bushes, and remove the driven joint. If worn or damaged, the six rubber bushes can now be prised out and renewed as a set. When fitting the rubbers, their top faces should be 5.5 – 6.5 mm (0.22 – 0.26 in) below the top edge of their individual bores.

4   Reassemble using the reverse of the dismantling procedure. Note that when refitting the driven joint, apply graphite grease to the 'bolt' ends which locate in the rubbers, and coat the internal central splines with a high melting point grease. Note also the O-ring around the central spigot base, behind the driven joint. Renew the O-ring if its condition is at all suspect. Ensure the tab washers securing the six driven joint retaining bolts, are bent over correctly.

15.3a Remove the six smaller bolts, with the locking washers straightened and ...

15.3b ... remove the driven plate and headed stubs – GN type shown

15.3c Prise out the rubbers if damaged or worn

15.4 Note O-ring around central spigot base

16 Rear wheel bearings: removal and replacement

1   The rear wheel assembly has two journal ball bearings; one bearing lying each side of the wheel hub. The procedure for the removal and examination of the rear wheel bearings is similar to that given for the front wheel bearings in Section 8 of this Chapter.

2   The rear wheel must be removed from the machine (refer to Section 10 of this Chapter) before access can be gained to the bearings. Commence by drifting out the right-hand bearing, using the same method as described for the front wheel. Before the driven joint side bearing is tapped out, the hollow central spacer should be removed, followed by the double diameter left-hand spacr.

3   When refitting the bearings, ensure they are inserted squarely and so will not tie or bind as they are drifted into position. Note the O-ring around the base of the central hub spigot. Renew the O-ring if its condition is at all suspect.

16.2a With RH bearing drifted out ...

16.2b ... remove the hollow central spacer and ...

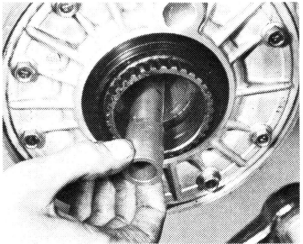

16.2c ... the double diameter LH spacer

17 Tyres: removal and replacement

1   At some time or other the need will arise to remove and replace the tyres, either as a result of a puncture or because replacements are necessary to offset wear. To the inexperienced, tyre changing represents a formidable task, yet if a few simple rules are observed and the technique learned, the whole operation is surprisingly simple.

2   To remove the tyre from either wheel, first detach the wheel from the machine. Deflate the tyre by removing the valve insert and when it is fully deflated, push the bead from the tyre away from the wheel rim on both sides so that the bead enters the centre well of the rim. Remove the locking cap and push the tyre valve into the tyre itself.

3   Insert a tyre lever close to the valve and lever the edge of the tyre over the outside of the wheel rim. Very little force should be necessary; if resistance is encountered it is probably due to the fact that the tyre beads have not entered the well of the wheel rim all the way round the tyre.

4   Once the tyre has been edged over the wheel rim it is easy to work around the wheel rim so that the tyre is completely free on one side. At this stage, the inner tube can be removed.

5   Working from the other side of the wheel, ease the other edge of the tyre over the outside of the wheel rim that is furthest away. Continue to work around the rim until the tyre is free completely from the rim.

6   If a puncture has necessitated the removal of the tyre, reinflate the inner tube and immerse in a bowl of water to trace the source of the leak. Mark its position and deflate the tube. Dry the tube and clean the area around the puncture with a petrol soaked rag. When the surface has dried, apply rubber solution and allow this to dry before removing the backing from the patch and applying the patch to the surface.

7   It is best to use a patch of self-vulcanising type, which will form a very permanent repair. Note that it may be necessary to remove a protective covering from the top surface of the patch, after it has sealed into position. Inner tubes made from synthetic rubber may require a special type of patch and adhesive, if a satisfactory bond is to be achieved.

8   Before refitting the tyre, check the inside to make sure that the agent which caused the puncture is not trapped. Check the outside of the tyre, particularly the tread area, to make sure nothing is trapped that may cause a further puncture.

9   If the inner tube has been patched on a number of past occasions, or if there is a tear or large hole, it is preferable to discard it and fit a new one. Sudden deflation may cause an accident, particularly if it occurs with the front wheel.

10   To replace the tyre, inflate the inner tube sufficiently for it to assume a circular shape but only just. Then push it into the tyre so that it is enclosed completely. Lay the tyre on the wheel at an angle and insert the valve through the rim tape and the hole in the wheel rim. Attach the locking cap on the first few threads, sufficient to hold the valve captive in its correct location.

11   Starting at the point furthest from the valve, push the tyre bead over the edge of the wheel rim until it is located in the central well. Continue to work around the tyre in this fashion until the whole of one side of the tyre is on the rim. It may be necessary to use a tyre lever during the final stages.

12   Make sure that there is no pull on the tyre valve and again commencing with the area furthest from the valve, ease the other bead of the tyre over the edge of the rim. Finish with the area close to the valve, pushing the valve up into the tyre until the locking cap touches the rim. This will ensure the inner tube is not trapped when the last section of the bead is edged over the rim with a tyre lever.

13   Check that the inner tube is not trapped at any point. Reinflate the inner tube, and check that the tyre is seating correctly around the wheel rim. There should be a thin rib moulded around the wall of the tyre on both sides, which should be equidistant from the wheel rim at all points. If the tyre is unevenly located on the rim, try bouncing the wheel when the tyre is at the recommended pressure. It is probable that one of the beads has not pulled clear of the centre well.

14   Always run the tyres at the recommended pressures and never under or over-inflate. The correct pressures for all uses are given in the Specifications Section of this Chapter.

15   Tyre replacement is aided by dusting the side walls, particularly in the vicinity of the beads, with a liberal coating of french chalk. Washing-up liquid can also be used to good effect.

18 Valve cores

1   Valve cores seldom give trouble, but do not last indefinitely. Dirt under the seating will cause a puzzling 'slow-puncture'. Check that they are not leaking by applying spittle to the end of the valve and watching for air bubbles.

19 Tyre valve dust caps

1   Tyre valve dust caps are often left off when a tyre has been replaced, despite the fact that they serve an important two-fold function. Firstly they prevent dirt or other foreign matter from entering the valve and causing the valve to stick open when the tyre pump is next applied. Secondly, they form an effective second seal so that in the event of the tyre valve sticking, air will not be lost.

2   Isolated cases of sudden deflation at high speeds have been traced to the omission of the dust cap. Centrifugal force has tended to lift the tyre valve off its seating and because the dust cap is missing, there has been no second-seal. Racing inner tubes contain provision for this happening because the valve inserts are fitted with stronger springs, but standard inner tubes do not, hence the need for the dust cap.

3   Note that when a dust cap is fitted for the first time, the wheel may have to be rebalanced.

20 Front wheel: balancing

1   It is customary on all high performance machines to balance the front wheel complete with tyre and tube. The out of balance forces which exist are eliminated and the handling of the machine is improved in consequence. A wheel which is badly out of balance produces through the steering a most unpleasant hammering effect at high speeds.

2   Some tyres have a balance mark on the sidewall, usually in the form of a coloured dot. This mark must be in line with the tyre valve, when the tyre is fitted to the inner tube. Even then, the wheel may require the addition of balance weights, to offset the weight of the tyre valve itself.

3   If the front wheel is raised clear of the ground and is spun, it will probably come to rest with the tyre valve or the heaviest part downward and will always come to rest in the same position. Balance weights must be added to a point diametrically opposite this heavy spot until the wheel will come to rest in ANY position after it is spun.

4   The cast-alloy wheels fitted to the GS 850 models have two weights of wheel balancing weight available. These balance weights are available in 20 gm (0.04 lb) and 30 gm (0.07 lb) sizes, which clip to the rim of the wheel. The relevant Suzuki part Nos are 55411-45100 (20 gm) and 55412-45100 (30 gm).

5   The remarks relating to the tyre balancing mark, and also to balancing in general, apply equally to the rear tyre. Because of the drag of the final drive components the tyre must be balanced with the wheel off the machine, supported on a suitable spindle.

20.4 Balance weights are available in two weights

21 Fault diagnosis: wheels, brakes and tyres

Symptom	Cause	Remedy
Handlebars oscillate at low speed	Buckled front rim Tyre not straight on rim	Check rim alignment by spinning the wheel. Correct by renewing the wheel. Check tyre alignment.
Machine lacks power and accelerates poorly	Brakes binding due to a wrongly adjusted caliper	Check brake pads and whether piston(s) sticking. Readjust caliper.
Brakes feel spongy	Air in hydraulic line Fluid leak in system	Bleed brakes. Replace faulty part.
Brakes grab when applied gently	Faulty caliper Warped disc	Replace with a new caliper. Replace disc/have disc skimmed.
Brake squeal	Glazed pads	Lightly sand the pads, and use the brake gently for a hundred miles or so until they have a chance to bed in properly.
	Extremely dirty and dusty front brake caliper and disc assembly	Clean with water; do not use high pressure spray equipment.
Excessive lever or pedal travel	Air in system, or leak in master cylinder or caliper; worn disc pads	Bleed the brake. Renew the cylinder seals. Renew the pads.
Brake pull-off sluggish	Sticking pistons in brake caliper	Overhaul caliper unit.
Tyres wear more rapidly in middle of tread	Over-inflation	Check pressures and run at recommended pressures
Tyres wear more rapidly at outer edges of tread	Under-inflation	Check pressures and run at recommended pressures.

Chapter 6 Electrical system

Refer to Chapter 7 for information relating to the 1981 to 1988 models

Contents

Specifications

Battery

Make	Yuasa
Type	12N14-3A/YB14L-A2
Voltage	12 volts
Capacity	14 Ah
Earth	Negative

Alternator

Make	Nippon Denso
Type	Permanent magnet rotor, 12-coil (GN model) 18-coil (GT/GLT model) stator
No load voltage @ 5000 rpm	More than 75V (GN model) or 80V (GT, GLT model)
Regulated voltage	14.0-15.5V @ 5000 rpm

Starter motor

Make	Mitsuba
Brush length	12-13 mm (0.47-0.51 in)
Service limit	6 mm (0.24 in)
Commutator undercut	0.6 mm (0.02 in)
Service limit	0.2 mm (0.008 in)
Relay resistance – yellow/green to earth	3 – 4 ohm

Bulbs

Headlamp:	
GS 850 GN (US)	50/40W Sealed beam
GS 850 GN (UK)	45/45W
GS 850 GT/GLT	60/55W QH
Pilot lamp (UK only)	4W
Tail/stop lamp:	
US models	8/23W (3/32cp)
UK models	5/21W
Flashing indicator lamps:	
US models	23W (32cp)
UK models	21W
Licence plate lamp – GN and GT models:	
US models	8W (4cp)
UK models	5W

Instrument lights	3.4W
Neutral indicator light	3.4W
Oil pressure warning light	3.4W
High beam indicator light	3.4W
Indicator warning light	3.4W
Fuel gauge light	3.4W (GLT model only)
Digital gear position lights (5)	3.4W (GLT model only)

All bulbs rated at 12 volts

Fuses

Headlamp	10A
Flashing indicator lamps	10A
Ignition	10A
Main	15A
Accessory output terminal	10A

1   General description

The Suzuki GS 850 models are fitted with a 12 volt electrical system powered by an alternator mounted on the left-hand end of the crankshaft. The alternator produces ac current and is of the three-phase type having a permanent magnet rotor and either a twelve-coil stator (GN models) or an eighteen-coil stator (GT/GLT models).

The alternating current ac produced by the alternator is converted to direct current (dc) by a full-wave silicon rectifier and is controlled to meet the voltage demands of the system by a solid state regulator. These two functions are performed by a combined regulator/rectifier unit.

On GN models and the GT model imported into the UK, only two of the three output phases are utilised during normal daylight running. The third phase output passes into the lighting switch and is brought into circuit only when the main lights are operating. The remaining models do not have this switching arrangement; the full three-phase output is applied through the rectifier/regulator at all times.

2   Testing the electrical system: general

1   Checking the electrical output and the performance of the various components within the charging system requires the use of test equipment of the multi-meter type. When carrying out checks, care must be taken to follow the procedures laid down and so prevent inadvertent incorrect connections or short circuits. Irreparable damage to individual components may result if reversal of current or shorting occurs. It is advised that unless some previous experience has been gained in auto-electrical testing, the machine may be returned to a Suzuki Service Agent or auto-electrician, who will be qualified to carry out the work and have the necessary test equipment.

2   If the performance of the charging system is suspect, the system as a whole should be checked first, followed by testing of the individual components to isolate the fault. The three main components are the alternator, the rectifier, and the regulator. The two last items are combined into a single unit. Before commencing the tests, ensure that the battery is fully charged as described in Section 7.

3   Charging system: checking the output

1st test

1   The first test to be performed will determine the direct output of the alternator prior to the output being passed through the regulator/rectifier unit. An ac voltmeter with a 0-100 volt range or a multimeter set to a similar ac voltmeter range will be required.

2   Detach the left-hand frame side panel and disconnect the leads from the alternator at the three snap connectors. Start the engine and measure the ac voltage between the white/blue wire and the yellow wire, with the engine running at 5000 rpm.

Repeat the test across the white/blue and white/green wires, and again across the white/green wire and the yellow wire. In each case the voltage must be at least 75 volts (GN model) or 80 volts (GT/GLT models). If the voltage is any less there is evidence that the alternator is at fault. To substantiate that a fault exists in the alternator a further, static, test may be made as described in Section 4.

3   If results of the first test are satisfactory, a second test may be made, this time checking the output subjected to the effects of the rectifier/regulator.

2nd test

4   Reconnect the alternator leads and double check that the battery is fully charged. If the specific gravity of the electrolyte is less than 1.260, recharge the battery before making the test. Remove the two centre fuses (GN models) or three upper fuses (GT/GLT models) from the bank of fuses fitted below the left-hand side panel. This will ensure that the charging system output is checked with the alternator working in no-load conditions. Connect a dc voltmeter with a 0-20 volt range, or a multimeter set to a suitable scale, across the battery terminals. The meter positive (+) probe should be connected to the battery positive terminal, and the meter negative (-) probe connected to the battery negative terminal.

5   Start the engine and check the output at 5000 rpm. If the regulated voltage is within the range 14/15.5 volts, the regulator/rectifier unit is functioning satisfactorily. A reading outside this range indicates that the unit may be at fault. To prevent damage to the alternator stator, this test must be accomplished with the engine running at 5000 rpm for the very least possible time. A further check on the regulator/rectifier unit, may be carried out as described in Section 5.

4   Alternator: testing

1   If after carrying out the first test in Section 3 of this Chapter it was found that the alternator was not functioning correctly, as indicated by the voltage reading, the alternator stator should be removed from the machine and tested for resistance using a multi-meter set to the resistance function.

2   Disconnect the leads from the alternator at the three snap connectors and remove the engine left-hand casing. The alternator stator is fitted to the inside of the casing.

3   Measure the resistance between the three leads (yellow, white/blue, white/green) in pairs. In all three tests, continuity (or low resistance) should be found. If non-continuity (or infinite resistance) is found between any two wires it is evidence that an open circuit has occurred in the stator coils. Make a further check between each lead and the stator core to check that the stator windings are isolated from the core. In each case a reading of non-continuity should be experienced. A measurable low resistance indicates that a short circuit has occurred and that during operation current will be running to earth.

4   If failure of the stator windings does occur, repair is unlikely to be practicable because the coils are sealed and essentially inseparable. If, however, the damage is external and easily visible the advice of an auto-electrician should be sought before consigning this component to the scrap bin.

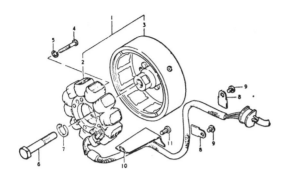

Fig. 6.1 Alternator – GS 850 GN model

1 Alternator assembly	7 Spring washer
2 Stator	8 Cable clamp – 2 off
3 Rotor	9 Screw – 2 off
4 Screw – 3 off	10 Bracket
5 Spring washer – 3 off	11 Screw
6 Bolt	

4.1 The combined regulator/rectifier is the finned component below the battery carrier

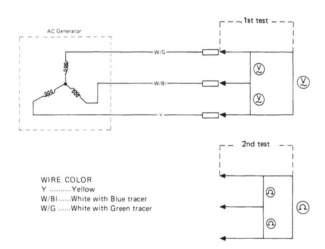

WIRE COLOR
Y ....Yellow
W/Bl......White with Blue tracer
W/G ......White with Green tracer

Engine speed	AC line voltage	Electrical condition of AC generator
5 000 r/min.	75V or over	AC generator is satisfactory
	75V or under	AC generator is faulty

Fig. 6.2 Alternator voltage and resistance test circuit

5 Regulator/rectifier: testing

1 If, after carrying out the second test in Section 3 of this Chapter there is evidence that the regulator/rectifier unit is malfunctioning a resistance test may be made to verify whether this is the case.

2 Disconnect the leads from the rectifier/regulator unit at the snap connectors and remove the bolt which secures the black/white earth lead. Measure the resistance between each lead as shown in the table in Fig. 6.3

3 If any one test does not conform with the expected result it is an indication that one or more internal diode in the regulator/rectifier unit has malfunctioned. Unfortunately this component is a sealed unit and cannot be repaired; a new one must be fitted if a malfunction occurs.

6 Battery: examination and maintenance

1 The GS 850 models are fitted with a 12 volt 14Ah battery.

2 The transparent plastic case of the battery permits the upper and lower levels of the electrolyte to be observed without disturbing the battery, by removing the left-hand side cover. Maintenance is normally limited to keeping the electrolyte level between the prescribed upper and lower limits and making sure that the vent tube is not blocked. The lead plates and their separators are also visible through the transparent case, a further guide to the general condition of the battery.

3 Unless acid is spilt, as may occur if the machine falls over, the electrolyte should always be topped up with distilled water to restore the correct level. If acid is spilt onto any part of the machine. It should be neutralised with an alkali such as washing

soda or baking powder and washed away with plenty of water, otherwise serious corrosion will occur. Top up with sulphuric acid of the correct specific gravity (1.260 to 1.280) only when spillage has occurred. Check that the vent pipe is well clear of the frame or any of the other cycle parts.

4   It is seldom practicable to repair a cracked battery case because the acid present in the joint will prevent the formation of an effective seal. It is always best to renew a cracked battery, especially in view of the corrosion which will be caused if the acid continues to leak.

5   If the machine is not used for a period, it is advisable to remove the battery and give it a refresher charge every six weeks or so from a battery charger. If the battery is permitted to discharge completely, the plates will sulphate and render the battery useless.

6   Occasionally, check the condition of the battery terminals to ensure that corrosion is not taking place and that the electrical connections are tight. If corrosion has occurred, it should be cleaned away by scraping with a knife and then using emery cloth to remove the final traces. Remake the electrical connections whilst the joint is still clean, then smear the assembly with petroleum jelly (NOT grease) to prevent recurrence of the corrosion. Badly corroded connections can have a high electrical resistance and may give the impression of a complete battery failure.

5.1 Alternator stator secured by three screws inside LH crankcase cover (GN model shown)

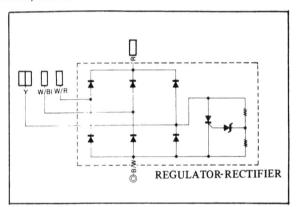

REGULATOR-RECTIFIER

Unit: Ω

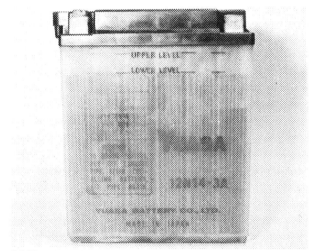
6.2 The electrolyte level must not be allowed to fall below the level shown

	⊕ Probe of tester				
		R	W/Bl	W/R	Y	B/W
⊖ Probe of tester	R		∞	∞	∞	∞
	W/Bl	5–7		∞	∞	∞
	W/R	5–7	∞		∞	∞
	Y	5–7	∞	∞		∞
	B/W	30 – 40	5–7	5–7	5–7	

Fig. 6.3 Rectifier/regulator test circuit diagram and test table

Wiring colour key:-

R   – *Red*
Y   – *Yellow*
B/W – *Black with white tracer*
W/Bl – *White with blue tracer*
W/R – *White with red tracer*

7   Battery: charging procedure

1   The normal charging rate for batteries is always one tenth of the battery's rated capacity. The 14 amp hour (Ah) battery fitted to the GS 850 range will therefore, require charging at 1.4 amps. It is permissible to charge at a more rapid rate in an emergency, but this shortens the life of the battery, and should be avoided. Always remove the vent caps when recharging a battery, otherwise the gas created within the battery when charging takes place may burst the case with disastrous consequences.

2   Make sure the charger connections to the battery are correct; red to positive and black to negative. It is preferable to remove the battery from the machine during the charging operation.

3   When the battery is reconnected to the machine, it is again important that the two leads are replaced on the correct terminals. If the leads are inadvertently reversed, the electrical system will be damaged permanently. The rectifier will be destroyed by a reversal of the current flow.

8 Fuses: location and replacement

1   A bank of fuses is contained within a small plastic box located on the electrical components mounting plate. To gain access to the fuse box, removal of the left-hand side frame cover is necessary. The front section of the fuse box pulls away from the base section leaving either four fuses visible (GN model) or five fuses visible (GT/GLT models). The reason for the variation in the number of fuses within the fuse box, is due to a slight redesigning of the electrical components mounting plates on the later models. On the GN model, a separate fused output terminal is provided, for electrical accessories to be fitted to the machine. On the GT/GLT models, the output terminal and its fuse have been combined within the fuse box. The inside of the fuse box front section contains spare fuses of each rating used in the system. The fuses used are four (including the output terminal) at 10A and one at 15A.

2   The main fuse, 15A, protects all the electrical systems. The three 10A fuses all protect specific circuits. One protects the headlamp, tail lamp, rear number plate lamp, instrument lamps and high beam indicator lamp. The second 10A fuse protects the brake lamp, flashing indicators and warning lamp repeater, and the horn. The third 10A fuse is responsible for protecting the electric start system and the ignition system.

3   Before replacing a fuse that has blown, check for the cause of the short. Check that no obvious short circuit has occurred. This will involve checking the electrical circuit to correct the fault. If this rule is not observed, the fuse will almost certainly blow again.

4   When a fuse blows while the machine is in operation, and no spare is available, a 'get you home' remedy is to remove the blown fuse and wrap it in silver paper before replacing it in the fuse holder. The silver paper will restore the electrical continuity by bridging the broken fuse wire. This expedient should **NEVER** be used if there is evidence of a short circuit or other major electrical fault, otherwise more serious damage will be caused. Replace the 'doctored' fuse at the earliest possible opportunity to restore full circuit protection.

9 Starter motor: removal, examination and replacement

1   An electric starter motor, operated from a small push-button on the right-hand side of the handlebars, provides either an alternative method of starting, without having to use the kickstart (GN model), or the sole method of starting the engine (GT/GLT models). The starter motor is mounted within a compartment at the rear of the cylinder block, closed by an oblong, chromium plated cover. Current is supplied from the battery via a heavy duty solenoid switch and a cable capable of carrying the very high current demanded by the starter motor on the initial start-up.

2   The starter motor drives a free running clutch immediately behind the generator rotor. The clutch ensures the starter motor drive is disconnected from the primary transmission immediately the engine starts. It operates on the centrifugal principle; spring loaded rollers take up the drive until the centrifugal force of the rotating engine overcomes their resistance and the drive is automatically disconnected.

3   To remove the starter motor from the engine unit, first disconnect the positive lead from the battery, to isolate the electrical system. Remove the cover plate which encloses the starter motor and detach the heavy duty cable from the terminal on the starter motor body. Temporarily detach the oil pressure switch lead from the switch. The starter motor is secured to the crankcase by two bolts which pass through the left-hand end of the motor casing. When these bolts are withdrawn, the motor can be prised out of position and lifted out of its compartment.

4   The parts of the starter motor most likely to require attention are the brushes. The end cover is retained by the two long screws which pass through the lugs cast on both end pieces. If the screws are withdrawn, the end cover can be lifted away and the brush gear exposed.

5   Lift up the spring clips which bear on the end of each brush and remove the brushes from their holders. The standard length and wear limit of the brushes is as follows:

standard length	service limit
12-13 mm (0.47-0.51 in)	6 mm (0.24 in)

6   Before the brushes are replaced, make sure the commutator is clean, on which they bear. Clean with a strip of fine glass paper pressed against the commutator whilst the latter is revolved by hand. Emery paper should **not** be used because abrasive fragments may embed themselves in the soft metal of the commutator and cause excessive wear of the brushes. Finish off the commutator with metal polish to give a smooth surface and finally wipe the segments over with a methylated spirits soaked rag to ensure a grease free surface. Check that the mica insulators, which lie between the segments of the commutator, are undercut. The standard groove depth is 0.6 mm (0.02 in) but if the average groove depth is less than 0.2 mm (0.008 in) the armature should be renewed or returned to a Suzuki Service Agent for re-cutting.

7   Replace the brushes in their holders and check that they slide quite freely. Make sure the brushes are replaced in their original positions because they will have worn to the profile of the commutator. Replace and tighten the end cover, then replace the starter motor and cable in the housing, tighten down

8.1a The fuse box and fused output terminal covers pull off to ...

8.1b ... facilitate renewal or examination

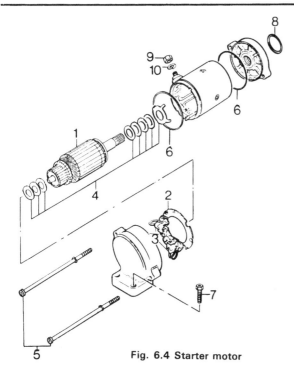

Fig. 6.4 Starter motor

1	Armature	6	O-ring – 2 off
2	Brush holder plate	7	Screw – 2 off
3	Brush – 2 off	8	O-ring
4	Shim – A/R	9	Nut
5	Bolt – 2 off	10	Washer

and re-make the electrical connection to the solenoid switch. Check that the starter motor functions correctly before replacing the compartment cover and sealing gasket.

10 Starter motor free running clutch: construction and renovation

1   Although a mechanical and not an electrical component, it is appropriate to include the free running clutch in this Chapter because it is an essential part of the electric starter system.

2   As mentioned in Chapter 1, the free running clutch is built into the alternator rotor assembly and will be found in the back of the rotor when the latter is removed from the left-hand end of the crankshaft. The only parts likely to require attention are the rollers and their springs, or the bush in the centre of the driven sprocket. Access to the rollers is gained by removing the three Allen bolts which retain the clutch body to the rear of the alternator rotor. Signs of wear or damage will be obvious and will necessitate renewal of the worn or damaged parts.

3   The bush in the centre of the driven sprocket behind the clutch will need renewal only after very extensive service.

4   To check whether the clutch is operating correctly, turn the driven sprocket clockwise. This should force the spring loaded rollers against the crankshaft and cause it to tighten on the crankshaft as the drive is taken up.

5   If the starter clutch has been dismantled, make sure the three Allen bolts are tightened fully with a small quantity of locking fluid applied to the bolt threads, to prevent their working loose.

11 Starter solenoid switch: function and location

1   The starter motor switch is designed to work on the electro-magnetic principle. When the starter motor button is depressed, current from the battery passes through windings in the switch solenoid and generates an electro-magnetic force which causes a set of contact points to close. Immediately the points close,

the starter motor is energised and a very heavy current is drawn from the battery. US models are fitted with a starter interlock system for safety; refer to Section 22.

2   This arrangement is used for at least two reasons. Firstly, the starter motor current is drawn only when the button is depressed and is cut off again when pressure on the button is released. This ensures minimum drainage on the battery. Secondly, if the battery is in a low state of charge, there will not be sufficient current to cause the solenid contacts to close. In consequence, it is not possible to place an excessive drain on the battery which, in some circumstances, can cause the plates to overheat and shed their coatings. If the starter will not operate, first suspect a discharged battery. This can be checked by trying the horn or switching on the lights. If this check shows the battery to be in good shape, suspect the starter switch which should come into action with a pronounced click. It is located on the electrical components mounting plate at the top, left-hand corner. It can be identified by the heavy duty starter cable connected to it. It is not possible to effect a satisfactory repair if the switch malfunctions; it must be renewed.

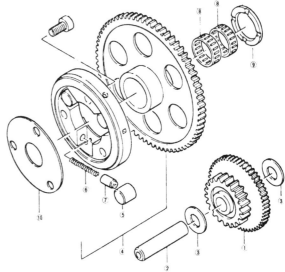

Fig. 6.5 Starter clutch

1	Idler gear	7	Plunger – 3 off
2	Idler gear shaft	8	Needle roller bearing –
3	Thrust washer – 2 off		2 off
4	Starter clutch assembly	9	Washer
5	Roller – 3 off	10	Shim
6	Spring – 3 off		

11.2 The starter motor solenoid switch

12 Headlamp: replacing bulbs and adjusting beam height

1   In order to gain access to the headlamp bulbs it is necessary first to remove the rim, complete with the reflector and headlamp glass. The rim is retained by three crosshead screws equally spaced around the headlamp shell. Remove the screws completely and draw the rim from the headlamp shell.

2   There are several variations in headlamp design and bulb types used throughout the GS 850 range. UK models of the GN type have a main headlamp bulb which is a push fit into the central bulb holder of the reflector. The bulb holder can be replaced in one position only to ensure the bulb is always correctly focussed. It is retained in position by notches on its edge which locate positively with lugs at the rear of the reflector. A bulb of the twin filament type, with a 45/45W rating is fitted.

3   The US market GN models are fitted with a sealed beam headlamp unit, rated at 50/40W. With this type of headlamp, if one filament blows, the complete unit must be renewed. To release the lamp unit, remove the horizontal adjusting screw and the three retaining screws and washers, from the collar that clamps the lamp shell to the headlamp rim. Make a note of the setting of the adjusting screw, otherwise it will be necessary to re-adjust the beam height after installing the new light unit. Remove the two screws to separate the sealed beam headlamp unit and the retaining ring from the rim, and remove a further three screws to release the sealed headlamp unit from the retaining ring. To reassemble the headlamp, reverse the dismantling procedure, noting that the marking TOP, faces upwards.

4   The GS 850 GT models, on both the UK and US markets, are fitted with headlamp units utilizing a quartz halogen type headlamp bulb. This bulb is rated at 60/55W. The QH bulb is a push fit in the central bulb holder of the reflector, and is retained in a manner similar to the bulb on the UK market GN model (see paragraph 2). There is, however, no separate bulb holder on these models, the bulb being secured by a rubber boot and a retaining spring. When removing the QH bulb, on no account should the quartz envelope be handled. Use a piece of clean, dry cloth to insulate the bulb from the hand when removal, or insertion, is necessary. If the bulb is to be discarded, obviously this precaution is not necessary. If the quartz glass is touched by the skin, a 'hot spot' will be formed. Due to the high temperatures developed during the working of these bulbs, this 'hot spot' will result in a weak point, and the bulb will fail prematurely.

5   On both the UK market models, a pilot lamp bulb is a standard fitting. The pilot lamp bulb is a bayonet fitting and fits within a small bulb holder which has the same type of attachment as the main bulb has to the reflector. This bulb has a rating of 4W. There are no pilot lamps fitted to US models.

6   The headlamp unit fitted to the GS 850 GLT model is slightly different again from those fitted to the rest of the range. The chromed unit appears initially to be a one-piece component. Removing two crosshead screws, one each side of the shell, however, allows the rim and headlamp internal unit to pull free from the shell. Rolling back the rubber boot reveals a retaining spring fitted to the rear of the bulb. Free the spring from the locating slot on the outside edge of the central bulb fitting hole, in the back of the reflector. Releasing the spring from tension enables the bulb to be removed. Refer to paragraph 4 for information on the quartz halogen bulb fitted to this headlamp.

7   Beam angle is adjusted by turning the adjusting screw provided. On the GN models (UK and US) the adjusting screw is in the 3 o'clock position when viewed from the front, and on the GT models (UK and US) the screw is in the 9 o'clock position, if viewed from the front. Turning the adjusting screw in or out results in the beam path being moved horizontally. The vertical beam height is adjusted manually. Loosen the left-hand and right-hand headlamp shell mounting bolts, and pivot the complete unit until the correct setting is achieved. Retighten the mounting bolts when height is correct. To adjust the horizontal beam angle on the GS 850 GLT model, a small crosshead screw is provided. This is fitted to the right-hand side of the headlamp rim at approximately the four o'clock position when the machine is viewed from the front three-quarters angle. Turn the adjusting screw clockwise or anti-clockwise in the same manner as for the other models. Vertical adjustment is as for the other models.

8   UK lighting regulations stipulate that the lighting system must be arranged so that the light will not dazzle a person standing at a distance greater than 25 feet from the lamp whose eye level is not less than 3 feet 6 inches above that plane. It is easy to approximate this setting by placing the machine 25 feet away from a wall, on a level road, and setting the beam height so that it is concentrated at the same height as the distance of the centre of the headlamp from the ground. The rider must be seated normally during this operation and also the pillion passenger, if one is carried regularly.

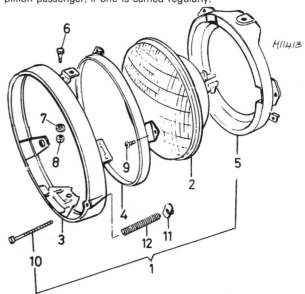

Fig. 6.6 Headlamp – GS 850 GN (US)

1 Headlamp unit	7 Washer – 2 off
2 Reflector unit	8 Nut – 2 off
3 Rim	9 Screw – 2 off
4 Inner rim	10 Beam adjusting screw
5 Mounting ring	11 Nut
6 Screw – 2 off	12 Spring

12.1 Three screws retain rim to reflector and glass (UK GN model shown)

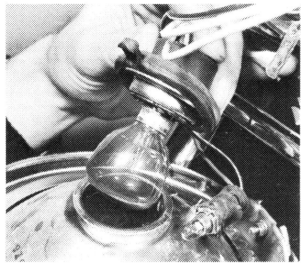

12.2 Main bulb holder and pilot bulb holder (UK GN model) both have rubber boots

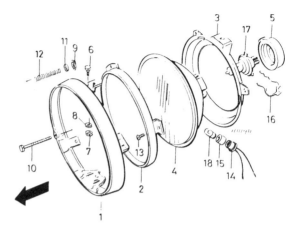

Fig. 6.7 Headlamp – GS 850 GT model

1 Rim	10 Beam adjusting screw
2 Inner rim	11 Washer
3 Back plate	12 Spring
4 Reflector unit	13 Screw
5 Grommet	14 Pilot bulb holder*
6 Bolt	15 Grommet*
7 Nut	16 Spring clip
8 Washer	17 Bulb
9 Nut	18 Pilot bulb*

* UK model only

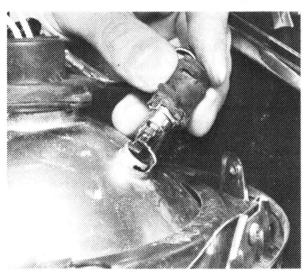

12.5 Pilot bulb holder has bayonet fixing as has the bulb (UK models)

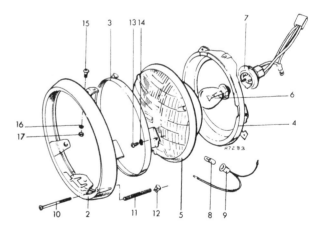

Fig. 6.8 Headlamp – GS 850 GN UK models

1 Spring washer – 3 off	10 Beam height adjusting
2 Rim	screw
3 Inner rim	11 Spring
4 Back plate	12 Nut
5 Reflector	13 Screw – 3 off
6 Bulb	14 Washer
7 Bulb holder	15 Screw
8 Pilot bulb	16 Washer
9 Pilot bulb holder	17 Nut

12.7 Use adjusting screw to maintain correct beam angle (GN model shown)

13 Stop and tail lamp: replacing the bulbs

1   The tail lamp unit on the GS 850 GN and GT models contains two bulbs. The larger of the two bulbs is of the twin filament type with, for the US market, a rating of 8/23W (3/32 cp) or for the UK market, a rating of 5/21W. The higher rated filament operates to give visible warning when the brake is applied, and the lower rated filament acts as a permanent warning lamp with the lights switched on. The second, single filament bulb (8W US models, 5W UK models) is used to illuminate the rear number plate. Stop lamp switches operate in conjunction with both the front and rear brakes in order that the system meets the statutory requirements of the country or state to which the machine is exported.

2   To gain access to the bulbs, unscrew the four crosshead screws which retain the plastic lens cover in position. The bulbs have a bayonet fitting and the tail/stop lamp has offset pins. This is so that the stop lamp filament cannot be inadvertently connected with the tail lamp and vice versa.

3   On the GLT model, there is only one tail/stop lamp bulb fitted, rated at 8/23W (3/32 cp). Due to the differing styling of the unit, a separate, number plate illuminating bulb, is not warranted. When replacing the bulb on this model, note that there are only two lens retaining screws, and they enter the unit from the rear, that is, the screw heads reside in the chromed backplate of the lamp unit.

14 Flashing indicator lamps: replacing bulbs

1   Flashing indicator lamps are fitted to the front and rear of the machine. They are mounted on short stalks through which the wires pass. Access to each bulb is gained by removing the two screws holding the plastic lens cover. The bulbs are of 23W (32 cp) rating on the US models and 21W rating on the UK market models.

15 Flasher relay unit: location and replacement

1   The flasher relay unit is located behind the left-hand side panel on the electrical components mounting plate, to the bottom right of the plate. The relay unit is supported on anti-vibration mountings made of rubber attached to the mounting plate.

2   If the flasher unit is functioning correctly, a series of audible clicks will be heard when the indicator lamps are in operation. If the unit malfunctions and all the bulbs are in working order, the usual symptom is one initial flash before the unit goes dead; it will be necessary to replace the unit complete if the fault cannot be attributed to any other cause.

3   In addition to the flasher relay unit, a self-cancelling unit is fitted. The electronic self-cancelling unit is incorporated in the indicator system. The unit automatically turn the indicators off, a certain time after the indicator switch has been operated. The time lapse is dependent upon the speed of the machine. If the machine is travelling faster than 15 km/h (9 mph), the unit cancels the indicators after approximately 9 seconds. When the machine is stationary, or running at a speed of less than 15 km/h (9 mph), the unit does not come into operation; the indicators continue to function until either cancelled manually, or the machine speed increases. The system may be overridden manually, for example, when overtaking a single vehicle when the full 9 seconds or automatically controlled operation would not be required, by simple depressing the switch downwards.

4   The self-cancelling unit is fitted behind the right-hand frame side cover, behind and to the left of the rear brake master cylinder. It is retained in position by a rubber retaining strap.

5   Take great care when handling either unit because they are easily damaged if dropped or subjected to rough treatment.

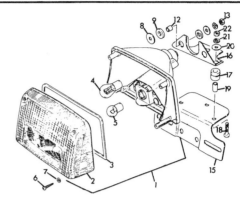

Fig. 6.9 Rear lamp – GS 850 GN model

1	Rear lamp assembly	13	Nut – 2 off
2	Reflector lens	14	Spring washer – 2 off
3	Rubber seal	15	Number plate mounting
4	Bulb		bracket
5	Bulb	16	Rear lamp mounting
6	Screw – 4 off		bracket
7	Sealing washer – 4 off	17	Grommet – 3 off
8	Washer – 2 off	18	Bolt – 3 off
9	Grommet – 2 off	19	Spacer – 3 off
10	Grommet – 2 off	20	Washer – 3 off
11	Washer – 2 off	21	Spring washer – 3 off
12	Spacer – 2 off	22	Nut – 3 off

16 Speedometer and tachometer: replacement of bulbs

1   Bulbs fitted to each instrument illuminate the dials when the main lights are in operation. The four bulbs fitted to the two instruments all have the same type of push fit bulb holder and are all of the small bayonet type fitting.

2   Access to the bulbs and holders is gained by removing the nuts from the studs on the bottom of each instrument which retain the chromed cover.

17 Indicator panel lamps: replacement of bulbs

1   An indicator lamp panel, advising the operator of the functioning of certain components, is fitted between the speedometer and tachometer heads. The panel contains five warning bulbs on all the models except the GS 850 GLT model which contains four. To gain access to the bulbs, remove the four screws which pass through the upper cover and lift the cover away. Each bulb is fitted to a separate bulb holder.

18 Gear selector indicator: location and operation

1   A gear selection indicator is fitted to all the GS 850 models, in addition to the neutral warning light. In this system, a switch on the change drum illuminates a digital display unit fitted in a console mounted between the speedometer and tachometer heads. The display unit indicates which of the gears in turn has been selected, by illuminating separate bulbs behind a numbered panel. If the unit malfunctions, removal of the warning lamp console cover will facilitate inspection of the bulbs and their holders. Each bulb can be replaced individually.

19 Horn: location and examination

1   The horn on the GS 850 models consists of two separate units each flexibly mounted on a steel strip which in turn are bolted to a mounting bracket. The mounting bracket is secured to the frame below the petrol tank by two bolts. With both the horn units being rubber mounted, a certain amount of flex will be evident even when the retaining bolts are fully tightened.

Care should be taken when refitting the horn units to ensure that they are positioned correctly, and do not come into contact with the underside of the petrol tank. When this occurs, with the machine in operation, a drumming sound will be produced, and a slight vibration may be felt through the petrol tank sides.

2   If one, or both, of the horn units malfunction, they must be replaced. It is a statutory requirement in the UK that the machine must be fitted with a horn in working order.

3   Both the horn units are adjustable by means of a small grub screw at the back of the body so that the volume can be varied if necessary. To adjust the volume, turn the screw about half a turn either way until the desired tone is required.

20 Ignition switch: removal and replacement

1   The combined ignition, lighting master switch and steering lock is mounted in a separate unit below the warning lamp and gear indicator consoles, forward of the handlebar cross-piece.

2   If the switch proves defective, it may be removed as a unit, after first removing the two retaining bolts and disconnecting the switch to the wiring loom, at the connector.

3   Refitting the switch can be accomplished by reversing the dismantling procedure. Repair of a defective switch is rarely practicable. It is preferable to purchase a new switch unit, which will probably necessitate the use of a different key.

14.1 Indicator bulbs have a bayonet fixing

15.1 Flasher unit relay is the black box shown here

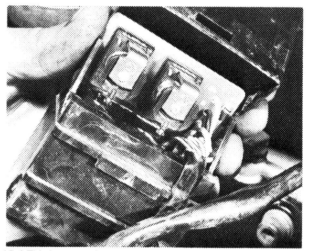
15.4 This box contains the self-cancelling system for the indicators

19.1 Ensure the horns are mounted correctly

20.1 Combined ignition, lighting master switch and steering lock is mounted below instruments

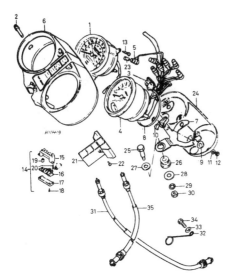

Fig. 6.10 Instrument console – GS 850 GN and GT models

1 Speedometer	18 Screw – 2 off
2 Tripmeter reset knob	19 Bulb – 5 off
3 Distance sensor	20 Screw – 2 off
4 Tachometer	21 Case
5 Bulb holder	22 Screw – 2 off
6 Instrument case	23 Bulb – 9 off
7 Mounting bracket	24 Bottom cover
8 Tachometer bracket	25 Bolt – 2 off
9 Mounting rubber – 4 off	26 Grommet – 2 off
10 Mounting rubber – 4 off	27 Plain washer – 2 off
11 Washer – 4 off	28 Plain washer
12 Nut – 4 off	29 Spring washer – 2 off
13 Screw – 4 off	30 Nut – 2 off
14 Gear position indicator	31 Speedometer cable
15 Casing	32 Cable guide
16 Bulb holder	33 Washer
17 Lower cover	34 Bolt
	35 Tachometer cable

21 Stop lamp switch: adjustment

1   All models have a stop lamp switch fitted to operate in conjunction with the front and rear brakes. The switch for the rear brake stop lamp, working from the rear brake pedal, is located immediately to the rear of the crankcase, on the right-hand side of the machine. It has a threaded body giving a range of adjustment.

2   If the stop lamp is late in operating, slacken the locknuts and turn the body of the switch anti-clockwise so that the switch moves upwards from the bracket to which it is attached. When the adjustment seems correct, tighten the locknuts and test.

3   If the lamp operates too early, the locknuts should be slackened and the switch body turned clockwise so that it is lowered in relation to the mounting bracket. The operation of the lamp is partly personal preference, but as a guide, it should operate after the brake pedal has been depressed by about $1\frac{1}{2}$ cm ($\frac{5}{8}$ inch).

4   A stop lamp switch is also incorporated in the front brake system. The mechanical switch is a push fit in the handlebar lever stock. If the switch timing needs to be adjusted, this can be accomplished by loosening the two screws that retain the switch to the body, and repositioning as necessary. If the switch malfunctions, repair is impracticable. The switch should be renewed.

22 Handlebar switches: general

1   In general, the switches give little trouble but, if necessary, they can be dismantled by separating the halves which form a split clamp around the handlebars. Note that the machine cannot be started with the ignition cut-out, on the right-hand switch unit, in the 'OFF' position; it must be moved to the 'RUN' position.

2   The GS 850 GN, GT and GLT models sold in the US, have been fitted wth an electric starter interlock system in addition to the ignition cut-out switch. This device interconnects the starter and the clutch handlebar lever; unless the clutch lever is operated, to disengage the clutch, the starter motor will not come into operation. The device operates by using a switch that interrupts the current passing from the starter motor button, until the clutch is disengaged.

3   Always disconnect the battery before removing any of the switches, to prevent the possibility of a short circuit. Most troubles are caused by dirty contacts, but in the event of the breakage of some internal part it will be necessary to renew the complete switch.

4   Because the internal components of each switch are very small and, therefore, difficult to dismantle and reassemble, it is suggested a special electrical contact cleaner be used to clean corroded contacts. This can be sprayed into each switch, without the need for dismantling.

23 Neutral indicator/gear position indicator switch: location and removal

1   A switch is incorporated in the gearbox which indicates, via a green lamp on the warning lamp console, when the neutral gear position has been selected. The switch also supplies information to the digital gear position indicator on the instrument console, indicating which of the five gears is engaged at any time. The switch is fitted into the left-hand side gearbox wall, roughly inset and below the secondary drive bevel gear unit.

2   In the event of failure, the switch may be unscrewed without draining the engine oil. It is retained by two screws. Disconnect the switch lead by unscrewing the single screw. Note that the switch can easily shear if excessive force is used when tightening the unit during refitting.

21.1 Stop lamp switch has threaded body for adjustment purposes

22.1a Switch clusters clamp around handlebar ends ...

22.1b ... and may be separated for maintenance (GN model shown)

24 Fuel gauge circuit: testing

1   If the fuel gauge appears to be faulty, the source of the fault is easily traced. Locate the fuel gauge sender leads. Separate the connectors, then join the harness side of the two leads with a length of wire. This eliminates the sender unit from the circuit, and if the ignition is now switched on, the fuel gauge should indicate full, indicating that the fault must lie in the sender unit. If the gauge does not respond, the fault lies in the instrument itself or its supply from the ignition switch. This can be checked using the relevant wiring diagram. A voltage regulator is incorporated in the gauge unit, to control the power supply more precisely; if either of these is faulty the gauge must be renewed. Note that on GN and GT models, this means renewing the complete tachometer (rev-counter).

2   Note that on later models the fuel gauge has a heavily-damped movement designed to leave the meter needle at its read-off value when the ignition is switched off. When checking operation, allow a few minutes for the gauge to respond.

3   If the fuel gauge sender is suspect, remove the fuel tank as described in Chapter 2. With the tank inverted on soft rag to protect the paintwork, remove the sender holding bolts and manoeuvre it out of the tank, taking care not to damage or bend the float. Check the sender resistances at the full and empty positions. Using a multimeter set it to its resistance function, measure the resistance across the sender Yellow/black and Black/white wires; in the full position there should be less than 1 – 5 ohm resistance, but in the empty position there should be 103 – 117 ohm resistance.

4   If the sender unit is faulty it must be renewed; repairs are not possible. If the circuit is merely inaccurate it is up to the owner to decide whether the expense of renewing the two components is justified; it is a simple matter to empty the tank and then refill it, noting for future reference the actual amount of fuel required to bring the meter needle to its various positions.

25 Fault diagnosis: electrical system

Symptom	Cause	Remedy
Complete electrical failure	Blown fuse	Check wiring and electrical components for short circuit before fitting new fuse.
	Isolated battery	Check battery connections, also whether connections show signs of corrosion.
Dim lights, horn and starter inoperative	Discharged battery	Remove battery and charge with battery charger. Check generator output and voltage regulator settings.
Constantly blowing bulbs	Vibration or poor earth connection	Check security of bulb holders. Check earth return connections.
Starter motor sluggish	Worn brushes	Remove starter motor Renew brushes
Parking lights dim rapidly	Battery will not hold charge	Renew battery at earliest opportunity
Flashing indicators do not operate	Blown bulb Damaged flasher unit	Renew bulb. Renew flasher unit.

The GS850 GG UK model

Chapter 7 The 1981 to 1988 models

Contents

Specifications

Information given below for the GS850 GN, GT and GLT models is supplementary to that given in the Specifications Sections of Chapters 1 to 6. Information applicable to all subsequent models is given only where different from their predecessors, ie the GS850 GT or GLT models. *Note: Where information is given under the heading 'All models' it applies to all models covered in this Manual.*

Model dimensions and weight

Overall length:
    GS850 GZ, GD, GE, GG .... 2195 mm (86.4 in)
    GS850 GLX, GLZ, GLD .... 2250 mm (88.6 in)

Overall width:
    GS850 GLX, GLZ, GLD .... 855 mm (33.7 in)
    All UK models .... 755 mm (29.7 in)

Overall height:
    GS850 GLX, GLZ, GLD .... 1130 mm (44.5 in)
    All UK models .... 1150 mm (45.3 in)

Ground clearance:
    GS850 GLX, GLZ, GLD .... 145 mm (5.7 in)
    GS850 GZ, GD, GE, GG .... 170 mm (6.7 in)

Dry weight:
    GS850 GLX, GLZ, GLD .... 536 lb (243 kg)
    GS850 GZ (UK), GE .... 536 lb (243 kg)
    GS850 GZ (US), GD .... 540 lb (245 kg)
    GS850 GG .... 582 lb (264 kg)

Gross vehicle weight rating – maximum permissible weight of machine, passenger(s), accessories and luggage:
    GS850 GN, GT, GLT, GLX .... 1003 lb (455 kg)
    GS850 GX, GZ, GD, GE, GG .... 1080 lb (490 kg)
    GS850 GLZ, GLD .... 1093 lb (496 kg)

Quick glance maintenance adjustments and capacities

Engine oil capacity – oil change only – GS850 GLX, GZ, GLZ, GD, GLD, GE, GG	3.0 lit (5.3 Imp pint, 3.1 US qt)
Front fork oil capacity – per leg:	
GS850 GZ (UK), GE, GG	217 cc (7.64 Imp fl oz)
GS850 GZ (US), GD	255 cc (8.62 US fl oz)
GS850 GLX, GLZ, GLD	245 cc (8.28 US fl oz)

	Normal riding	High speed riding
Tyre pressures – tyres cold:		
GS850 GLX, GLZ, GLD – pillion:		
Front	25 psi (1.75 kg/cm^2)	32 psi (2.25 kg/cm^2)
Rear	36 psi (2.50 kg/cm^2)	40 psi (2.80 kg/cm^2)
GS850 GG – solo, rear tyre	28 psi (2.00 kg/cm^2)	36 psi (2.50 kg/cm^2)

Recommended lubricants

Front forks:	
GS850 GX (UK)	1:1 mixture of Automatic Transmission Fluid (ATF) and SAE 10W/30 engine oil
All other later models	SAE 15 fork oil

Specifications relating to Chapter 1

Valves and springs

Valve collet groove to stem end minimum length – GS850 GD, GE, GG	4.4 mm (0.1732 in)

Clutch

Friction plate tongue width – all models:	
Standard	11.8 – 12.0 mm (0.4646 – 0.4724 in)
Minimum	11.0 mm (0.4331 in)
Clutch spring minimum free length – GS850 GX, GZ, GLZ, GD, GLD, GE, GG	38.4 mm (1.51 in)

Gearbox

Reduction ratios – GS850 GLX, GLZ, GLD:	
2nd gear	1.722:1 (31/18T)
5th gear	0.923:1 (24/26T)
Selector fork claw end to pinion groove clearance – all models:	
Standard	0.40 – 0.60 mm (0.0158 – 0.0236 in)
Maximum	0.80 mm (0.0315 in)
Selector fork claw end standard thickness – all models	4.95 – 5.05 mm (0.1949 – 0.1988 in)
Pinion groove standard width – all models	5.45 – 5.55 mm (0.2146 – 0.2185 in)

Torque wrench settings

	kgf m	lbf ft
Cylinder head cover bolts	0.6 – 1.0	4.5 – 7
Cylinder head:		
6 mm bolts	0.7 – 1.1	5 – 8
10 mm nuts	3.5 – 4.0	25.5 – 29
Camshaft cap bolts	0.8 – 1.2	6 – 8.5
Camshaft sprocket bolts:		
6 mm Allen bolts	0.8 – 1.2	6 – 8.5
6 mm hex-head bolts	1.5 – 2.0	11 – 14.5
7 mm hex-head bolts	2.4 – 2.6	17.5 – 19
Cam chain jockey sprocket bolts	0.6 – 1.0	4.5 – 7
Alternator rotor bolt	6.0 – 7.0	43.5 – 50.5
Crankcase bolts:		
6 mm bolts	0.6 – 1.0	4.5 – 7
8 mm bolts	1.8 – 2.2	13 – 16
Drive shaft bolts	3.0 – 4.0	21.5 – 29
ATU retaining bolt – GS850 GN, GT, GLT, GX, GLX, GLZ ..	1.8 – 2.3	13 – 16.5
Signal generator rotor bolt – GS850 GZ, GD, GLD, GE, GG	2.5 – 3.5	18 – 25.5
Engine mountings:		
8 mm nuts	N/Av	N/Av
10 mm nuts	2.5 – 4.5	18 – 32.5
12 mm nut	4.5 – 7.0	32.5 – 50.5

Specifications relating to Chapter 2

Fuel tank capacity – GS850 GLX, GLZ, GLD

Overall	17 lit (4.5 US gal)
Reserve	4.5 lit (1.2 US gal)

Carburettors

ID number:
GS850 GZ (UK), GE, GG   45170
GS850 GZ (US), GLZ, GD, GLD   45160
Pilot air jet – GS850 GZ (US), GLZ, GD, GLD   180
Pilot mixture screw – GS850 GX (UK), GZ (UK), GE, GG ...   Preset

Lubrication system

Engine/transmission oil capacity – GS850 GLX, GZ, GLZ,
GD, GLD, GE, GG:
  Oil change only – without filter ....   3.0 lit (5.3 Imp pint, 3.1 US qt)
Oil pump output pressure @ 3000 rpm – GS850 GZ, GD,
GLD, GE, GG ....   1.4 – 3.6 psi (0.10 – 0.25 kg/cm^2) @ oil temperature of
  60°C/140°F

Specifications relating to Chapter 3

Ignition HT coil – GS850 GZ, GD, GLD, GE, GG

Secondary winding resistance – Plug cap to Plug cap ....   30 – 40 K ohm approx

Signal generator – GS850 GZ, GD, GLD, GE, GG

Pick-up/pulser coil resistances – Blue to Yellow,
Black to Green ....   140 – 180 ohm

Specifications relating to Chapter 4

Front forks

Adjustment:
GS850 GZ (UK), GE, GG ....   4-way adjustable spring preload
GS850 GD ....   Air assisted, single valve (fork legs linked)

	Standard	Minimum
Fork spring free length:		
GS850 GZ (UK), GE, GG	N/Av	459.0 mm (18.0708 in)
GS850 GLX, GLZ, GLD	N/Av	516.0 mm (20.3149 in)

Fork oil quantity – per leg:
GS850 GZ (UK), GE, GG ....   217 cc (7.64 Imp fl oz)
GS850 GZ (US), GD ....   255 cc (8.62 US fl oz)
GS850 GLX, GLZ, GLD ....   245 cc (8.28 US fl oz)
Fork oil type:
GS850 GX (UK) ....   1:1 mixture of Automatic Transmission Fluid (ATF) and SAE
  10W/30 engine oil
All other later models ....   SAE 15 fork oil
Fork oil level – GS850 GLX, GLZ, GLD:
Leg fully compressed ....   260 mm (10.24 in)

Rear suspension

Travel – GS850 GZ, GD, GE, GG ....   110 mm (4.3 in)

Torque wrench settings

	kgf m	lbf ft
Drive shaft bolts	3.0 – 4.0	21.5 – 29
Steering stem head bolt – GS850 GZ, GLZ, GD, GLD, GE, GG	2.0 – 3.0	14.5 – 21.5

Specifications relating to Chapter 5

Wheels

	Front	Rear
Size:		
GS850 GN, GT, GLT, GX	MT 1.85 x 19	MT 2.50 x 17
GS850 GZ, GD, GE, GG	MT 2.15 x 19	MT 2.50 x 17
GS850 GLX, GLZ, GLD	MT 1.85 x 19	MT 2.75 x 16
Wheel spindle maximum runout	0.25 mm (0.01 in)	

Brakes

	GS850 GN	All other models
Rear brake pedal height – GS850 GLX, GLZ, GLD	15 mm (0.6 in)	
Brake master cylinder and caliper dimensions:		
Front master cylinder bore ID	15.870 mm (0.6248 in)	15.870 – 15.913 mm (0.6248 – 0.6265 in)
Front master cylinder piston OD	15.800 mm (0.6221 in)	15.827 – 15.854 mm (0.6231 – 0.6242 in)

Front brake caliper bore ID	42.850 mm (1.6870 in)	38.180 – 38.256 mm (1.5032 – 1.5061 in)
Front brake caliper piston OD	42.820 mm (1.6858 in)	38.098 – 38.148 mm (1.4999 – 1.5018 in)
Rear master cylinder bore ID	14.000 mm (0.5512 in)	14.000 – 14.043 mm (0.5512 – 0.5529 in)
Rear master cylinder piston OD	13.960 mm (0.5496 in)	13.957 – 13.984 mm (0.5495 – 0.5506 in)
Rear brake caliper bore ID	38.180 mm (1.5032 in)	38.180 – 38.256 mm (1.5032 – 1.5061 in)
Rear brake caliper piston OD	38.150 mm (1.5019 in)	38.098 – 38.148 mm (1.4999 – 1.5018 in)

Tyres
Type:

GS850 GN, GT, GLT, GX, GZ (UK) GE, GG .... Tubed

GS850 GLX, GZ (US), GLZ, GD, GLD .... Tubeless

Size:	**Front**	**Rear**
GS850 GLX, GLZ, GLD	100/90 x 19-57H	130/90 x 16-67H
All other models	3.50H x 19-4PR	4.50H x 17-4PR

Tyre pressures – tyres cold

GS850 GLX, GLZ, GLD – pillion:	**Normal riding**	**High speed riding**
Front	25 psi (1.75 kg/cm^2)	32 psi (2.25 kg/cm^2)
Rear	36 psi (2.50 kg/cm^2)	40 psi (2.80 kg/cm^2)
GS850 GG – solo, rear tyre	28 psi (2.00 kg/cm^2)	36 psi (2.50 kg/cm^2)

Torque wrench settings

	kgf m	**lbf ft**
Rear spindle nut – except GS850 GN	5.0 – 8.0	36 – 58
Brake hose banjo union bolts – except GS850 GN	1.3 – 1.8	9.5 – 13

Specifications relating to Chapter 6

Bulbs

Fuel gauge light – GS850 GLZ, GLD	12V, 1.7W
Side stand warning lamp – GS850 GZ, GLZ, GD, GLD, GE, GG	12V, 3.4W

1 Introduction

The preceding Chapters of this manual cover the Suzuki GS850 GN, GT and GLT models sold in the UK and US from 1979 to late 1980. This Chapter covers all subsequent models, describing only those features which require additional information or a modified working procedure.

When working on one of these later models, first use the notes given below to identify exactly the particular machine concerned, then check in this Chapter for any information relevant to the procedure which is to be carried out. If no information is given then the task will be substantially the same as that described in the relevant part of Chapters 1 to 6. *Note that Sections 2 and 3 of this Chapter also apply to the GS850 GN, GT and GLT models.*

Note: The Suzuki GS850 models, by normal Japanese standards, have remained largely unchanged throughout their production run. This does not mean, however, that components which are not mentioned here have remained unchanged. In many cases parts have been modified to a greater or lesser degree and may not be interchangeable with earlier versions. Great care is required when obtaining replacement parts to ensure that the correct item is fitted.

As with the earlier models, revised versions have appeared, which are identified at all times in this manual by their Suzuki production code. To assist the owner in identifying his or her machine, given below are the frame numbers with which each version's production run commenced, its approximate dates of import, and any significant identifying features.

GS850 GX (UK)

Frame numbers start at GS850-128084. Imported into the UK from December 1980 to March 1984, this model can be identified by black plastic square-bodied flashing indicator lamps and by the revised graphics on the fuel tank, side panels and seat tail cover.

GS850 GX (US)

Frame numbers start at JS1GS71G B2100001. Imported into the US from September 1980 to 1981, this model can be identified by the same features as the UK model described above.

GS850 GLX (US)

Frame numbers start at JS1GS71L B2100001. Imported into the US from September 1980 to 1981, this model can be identified by its modified headlamp brackets, by the two-tone paintwork on the fuel tank and side panels, and by the modified air filter side cover and side panel styling. The wheels are now highlighted in black (previously grey) and carry tubeless tyres; the rear wheel diameter is 16 inches. The footrests are now mounted on separate cast aluminium alloy plates which are bolted to each side of the frame and also carry the exhaust silencer mountings. Only one horn is now fitted.

GS850 GZ (UK)

Frame numbers start at GS850-131533. Imported into the UK from June 1983 to December 1985, this model can be identified by its extended pillion passenger grabrail and modified seat. The seat is no longer on hinged mountings but is now located on two hooks at the front and a locking catch at the rear, the lock being mounted in the more streamlined seat tail cover. Removal and refitting is now as described for the GLT model in Chapter 4.18. The paintwork and graphics are revised, as are the air filter side covers and the side panels. The wheels are now highlighted in black (formerly grey), the flashing indicator lamps are now chrome-plated, the choke is now operated by a lever mounted under the left-hand handlebar switch cluster and the front brake hydraulic fluid reservoir is now an integral part of the master cylinder body. The footrests and exhaust silencers are now bolted to separate cast alloy plates mounted on each side of the frame. A side stand warning lamp is incorporated in the instrument assembly.

GS850 GZ (US)

Frame numbers start at JS1GS72G C2100001. Imported into the US from November 1981 to 1982, this model's new identifying features are as for the UK model described above, with the addition of tubeless tyres.

GS850 GLZ (US)

Frame numbers start at JS1GS71L C2100001. Imported into the US from September 1981 to 1982, this model features two-tone paintwork and a cast aluminium alloy pillion passenger grabrail. The front brake master cylinder now incorporates an integral square-bodied fluid reservoir and the choke is now operated by a lever mounted under the left-hand handlebar switch cluster. A side stand warning lamp is incorporated in the instrument assembly.

GS850 GD (US)

Frame numbers start at JS1GS72G D2100001. Imported into the US from October 1982 to 1983, this model features a modified seat with a cast aluminium alloy pillion passenger grabrail and two-tone paintwork with black-painted crankcases, cylinder block and head. The seat tail cover now includes a storage box, the rear suspension units have small shrouds fitted over their upper ends, and a linking hose is fitted to connect the two front fork legs when setting their air pressure.

GS850 GLD (US)

Frame numbers start at JS1GS71L D2100001. Imported into the US from October 1982 to 1983, this model features square-bodied mirrors and flashing indicator lamps, black-painted crankcases, cylinder block and head, small shrouds over the upper ends of the rear suspension units and a small cover over the handlebar centre.

GS850 GE (UK)

Frame numbers start at GS850-132500. Imported into the UK from June 1985 to May 1986, this model features a cast aluminium alloy pillion passenger grabrail and black-painted crankcases, cylinder block and head. The seat tail cover now includes a storage box and the rear suspension units have small shrouds fitted over their upper ends.

GS850 GG (UK)

Frame numbers start at GS850-132895. Imported into the UK from May 1986 to December 1988, this model is identical to the GS850 GE apart from minor alterations to the instruments' warning lamp cluster, the flashing indicator lamps and the tail lamp.

---

2 Routine maintenance: revised service intervals – all models

---

Owners should note that the main service intervals have been slightly revised in the case of UK models, and extended in the case of US models. Carry out the tasks as described in the Routine maintenance section at the front of this manual, with reference as necessary to this chapter or to Chapters 1 to 6 for further information. Service intervals are as follows:

Daily pre-riding check

Check the steering for smoothness, free play and full movement
Check the brakes for correct lever or pedal play and check that there are no fluid leaks
Check the tyre pressures and tread depth, and that the tyres are undamaged

Check the engine, secondary drive and final drive oil levels. Add oil if necessary
Check the fuel level
Check all lights, flashing indicators, horn(s), warning lamps and instruments
Check the controls for correct adjustment, lubrication and operation and that the kill switch and stands operate correctly
Check that the suspension settings are correct (particularly the front fork air pressure on US models), and the suspension is operating properly

Monthly, or approximately every 500 miles (800 km)

Repeat the full daily check then carry out the following:
    Check the brake hydraulic fluid levels
    Check the battery electrolyte level, terminals and vent pipe

UK models – Every 1500 miles (2500 km) or three monthly
US models – Every 2000 miles (3000 km) or six monthly

Change the engine oil
Check the front fork air pressure – US models only

UK models – Every 3000 miles (5000 km) or six monthly
US models – Every 4000 miles (6000 km) or annually

Repeat all previous tasks, then carry out the following:
Change the engine oil and renew the oil filter element
Check the tightness of **all** nuts, bolts and fasteners, but particularly the cylinder head nuts and bolts and the exhaust system mountings
Clean the air filter element
Check the spark plugs
Check the valve clearances
Lubricate and adjust the throttle, choke, and clutch cables
Check the carburettors and engine idle speed
Check the brakes and lubricate the lever and pedal pivots
Check and lubricate the stand pivots
Check the steering head bearings and suspension components
US models – note that hydraulic brake fluid must be changed annually
The manufacturer recommends that the compression pressure and oil pressure be checked at this interval

UK models – Every 6000 miles (10 000 km) or annually
US models – Every 7500 miles (12 000 km) or two yearly

Repeat all previous tasks then carry out the following:
Renew the spark plugs
Clean the oil strainer (sump) filter
Change the secondary drive gear oil
Change the final drive gear oil
Grease the instrument drive cables and throttle twistgrip
Change the front fork oil
Grease the automatic timing unit – GN, GT, GLT, GX, GLX, GLZ models only
UK models – note that hydraulic brake fluid must be changed annually
US models – note that the fuel pipes and critical brake components must be renewed every two years

UK models – Every 12 000 miles (20 000 km) or two yearly
US models – Every 15 000 miles (24 000 km) or four yearly

Repeat all previous tasks then carry out the following:
Grease the steering head bearings
Grease the swinging arm pivot bearings
Grease the wheel bearings and the speedometer drive gearbox
UK models – note that the fuel pipes and critical brake components must be renewed every two years

Additional routine maintenance

The hydraulic brake fluid must be changed annually

The fuel feed pipes must be renewed every two years (four years, later models)

The brake hydraulic hoses must be renewed every two years (four years, later models)

The brake master cylinders and calipers must be overhauled and all seals renewed every two years

Note that the service intervals specified for the UK GS850 GG model are the same as those given above for US models. However, owners are advised that the revised UK model schedule given above is more realistic for machines in everyday use under UK conditions.

3 Carrying out a compression test – all models

1 A good indication of the degree of wear in the engine top end components can be gained by carrying out a compression test. This will require the use of an accurate compression tester (gauge) which has an adaptor suitable for use with 14 mm spark plug threads and is extended sufficiently to reach the deeply-recessed spark plug locations.

2 The engine must be fully warmed up to normal operating temperature after the valve clearances have been checked, and adjusted if necessary (see Routine maintenance), the cylinder head nuts and bolts must be correctly tightened and the battery must be in good condition and fully charged for the test results to be accurate.

3 Remove the spark plugs, fit them to the plug caps and lay the plugs on the cylinder head so that the plug metal bodies are earthed to the engine. This is to prevent damage to the ignition system; as a further precaution switch off the engine kill switch.

4 Attach the compression tester following its manufacturer's instructions. Check that the choke is in the fully-open position (for normal running) and switch on the ignition.

5 Open the throttle twistgrip fully and turn the engine over on the starter motor. The gauge reading should rapidly increase to a maximum level and settle at that (after about 4 – 7 seconds): note the reading and repeat the operation on the remaining cylinders.

6 If the pressures obtained are outside the specified tolerances the cylinder head must be removed for examination. Similarly if there is a marked discrepancy between any cylinders careful examination will be necessary; the maximum permissible difference is 2 kg/cm^2 (28 psi). In the unlikely event of the recorded pressures being above those specified, it is probable that excessive amounts of carbon have built up in the combustion chambers; these must be removed as soon as possible.

7 If the pressures are too low, pour a small quantity of oil into the combustion chamber(s) and repeat the test. If the pressures recorded increase significantly, then the piston, piston rings and cylinder bore are excessively worn (the oil having provided a temporary seal around the piston rings). This will require the removal of the cylinder head and block to find the cause and rectify it.

8 If the pressures do not alter after the oil has been added, then the cylinder head gasket or valves are leaking. These can be examined after the cylinder head has been removed, but the cylinder bores should be examined carefully (as far as possible) while the opportunity arises.

4 Engine/gearbox unit: modifications

Cylinder head cover bolts

1 GS850 GZ and GLZ (from engine number 155015 on), GD, GLD, GE and GG models are all fitted with a modified cylinder head cover, retained by nineteen bolts instead of the sixteen mentioned in Chapter 1. Note that this does not include the four shorter bolts which secure the breather cover on all models.

Tachometer drive

2 GS850 GD, GLD, GE and GG models are fitted with a slightly modified tachometer drive pinion housing; with reference to Fig. 1.1, note that items 25 and 29 are now a single component.

Camshaft sprocket mountings

3 If the camshaft sprockets are separated from the camshafts for any reason, note carefully the torque wrench settings (given in the Specifications Section of this Chapter) applicable to each type of fastener. Up to engine number 150525, all models used 6 mm (thread size) Allen bolts; after this, 7 mm (thread size) flanged, hexagon-headed bolts were used. Note that this means that early and late-type camshafts and sprockets cannot be interchanged. As interim type, using 6 mm (thread size) hexagon-headed bolts with a single tab washer securing each pair, may be found.

4 Whatever type of fastener is fitted, note that on reassembly the threads of the bolts and of the camshaft flanges must be thoroughly degreased and that a drop of thread-locking compound must be applied before the bolts are fitted and tightened to the correct torque setting. Suzuki recommend their own Thread lock Super 1303B (UK) or 1363A (US); if this is not locally available, be careful to use only an equivalent grade from a high-quality range such as Loctite.

Valve timing

5 While the basic procedure of setting the valve timing remains as described in Chapter 1.53, note that on GS850 GZ, GD, GLD, GE and GG models the crankshaft index marks are altered due to the fitting of a simple ignition rotor instead of the earlier ATU. Refer to the accompanying illustration for details.

6 Note that the crankshaft must be rotated until the 1.4 cylinder 'T' mark on the ignition rotor is aligned exactly with the metal pole protruding from the centre of the 1.4 cylinder (left-hand, looking at the signal generator plate) pulser/pick-up coil.

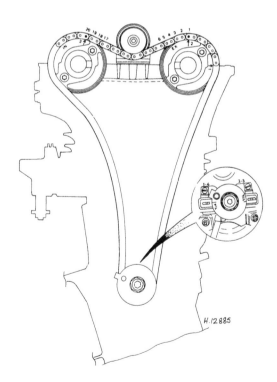

Fig. 7.1 Valve timing index marks – GS850 GZ, GD, GLD, GE and GG models

Cam chain tensioner

7   Note that the tensioner assembly oil seal is available separately on later models, also that an O-ring is fitted to the locking screw on some models. Check with a good Suzuki service agent for details of what can be obtained.

8   On GS850 GLD, GE and GG models the cam chain rear guide block (items 9, 10 and 11, Fig. 1.10) is no longer fitted.

Crankcase cover screws – GS850 GZ, GLZ, GD, GLD, GE and GG models

9   The screws securing the clutch cover, signal generator unit cover, alternator/starter motor cover and secondary drive unit cover are changed to the hexagon-headed type, of 7 mm or 8 mm thread size. These must be removed and refitted using slim-walled, deep, socket spanners of the appropriate size; these are usually only available in the smaller drive sizes, so ensure that all the necessary tools and adaptors are available before starting work. The Suzuki service tools are available under part numbers 09900-06711 (7 mm) or 09900-00302-015 (8 mm).

Separating the crankcase halves – GS850 GX and GLX models

10   This procedure is the same as described for the late GS850 GT and GLT models in Chapter 1, Sections 17 and 39, but note that the GS850 GLX model *only* is fitted with an oil feed nozzle and sealing O-ring set in the mating surface of the crankcase halves, towards the rear end. Check that the nozzle is clean whenever it is disturbed and always renew the sealing O-ring to prevent oil leaks and the risk of losing oil pressure.

Separating the crankcase halves – GS850 GZ, GLZ, GD, GLD, GE and GG models

11   With reference to Chapter 1, Sections 17 and 39, note that all these later models have crankcase halves which are clamped by a reduced number of bolts, and in a different pattern. Viewed from above (crankcases the correct way up on the workbench) there are seventeen bolts; seven 8 mm bolts (four of which are secured by nuts, two with washers) and ten 6 mm. Viewed from below (crankcases inverted to rest on the cylinder studs and suitable supports) there are twenty-one bolts; twelve 8 mm and nine 6 mm. Always ensure that the bolts are slackened or tightened evenly and progressively using the numerical sequence cast in the crankcases themselves. Check that *all* fasteners are removed before attempting to separate the crankcase halves and that all are refitted and correctly tightened on reassembly.

12   On all these later models note also the presence of an oil feed nozzle and sealing O-ring as described in paragraph 10.

Clutch release mechanism – except GS850 GX model

13   With reference to Fig. 1.4, note that on all later models (except the GS850 GX) the washer (item 23) is deleted.

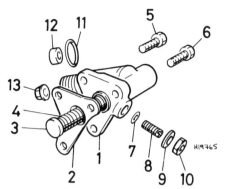

Fig. 7.2 Cam chain tensioner – later models

1	Tensioner body	8	Adjuster screw
2	Gasket	9	Washer
3	Plunger	10	Locknut
4	Spring	11	O-ring
5	Bolt	12	Oil seal
6	Bolt – 2 off	13	Nut
7	O-ring		

Clutch cover oil jet – except GS850 GX model

14   On all later models (except the GS850 GX) the clutch oil feed is controlled by a jet (size 110) which is screwed into the clutch cover gasket surface, at the rear of the release shaft boss. Whenever the cover is removed, check that the jet is clear; it can be unscrewed for cleaning if required.

Gear selector mechanism

15   With reference to Fig. 1.17, the two rear (layshaft) selector forks and the gearchange shaft return spring (items 1 and 26 respectively) are of a modified type on all later models. Removal and refitting are as described in Chapter 1.

---

5   Fuel tap: general – GS850 GLT, GLX, GZ, GLZ, GD, GLD, GE and GG models

---

1   With reference to the accompanying illustrations for details, note that some later models are fitted with modified fuel taps. All are of the same basic type and owners should note that no seals or washers are listed separately for any of them.

2   This means that if the tap develops a fuel leak it must be renewed to cure the fault. Similarly, if the tap becomes severely

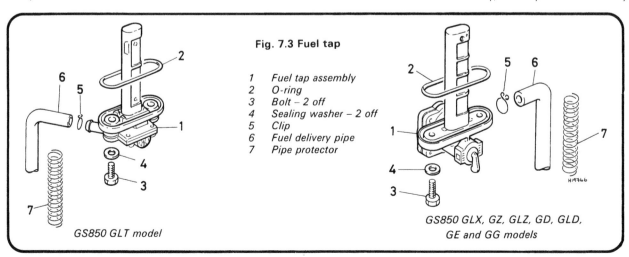

Fig. 7.3 Fuel tap

1	Fuel tap assembly
2	O-ring
3	Bolt – 2 off
4	Sealing washer – 2 off
5	Clip
6	Fuel delivery pipe
7	Pipe protector

GS850 GLT model

GS850 GLX, GZ, GLZ, GD, GLD, GE and GG models

blocked it is likely that the only effective long-term cure is to renew it; in some cases it may not be possible to dismantle the tap.

3 If the vacuum side of the tap is disturbed for any reason, note that the tap must be renewed as an assembly if the diaphragm is found to be split or perished. Check carefully how the small coil spring is located before removing it, and ensure that all components are refitted exactly as they are found; make notes and/or drawings as necessary to ensure this. Note particularly the small O-ring set around the diaphragm centre boss; ensure that this is not dislodged on refitting.

6 Carburettors and choke control: modifications – GS850 GZ, GLZ, GD, GLD, GE and GG models

1 The carburettors fitted to these later models are basically

similar to the constant depression type shown in Fig. 2.2, as fitted to the GS850 GT, GLT, GX and GLX models. There are, however, several modifications of varying significance, the most important of which are the changes to the throttle and choke cable mounting points and operating linkages. Refer to the accompanying illustration for details.

2 Note also that these later models are fitted with a choke which is operated by a handlebar-mounted lever. The lever is retained by a single central pivot screw and its movement is controlled by a wave washer. To dismantle the assembly remove the screw, disengage the lever from the cable end nipple and withdraw the various components. On reassembly grease the wave washer before installing it.

3 Note that since the choke cable no longer runs down through the steering stem, the steering stem nut is replaced by a bolt (see Section 11).

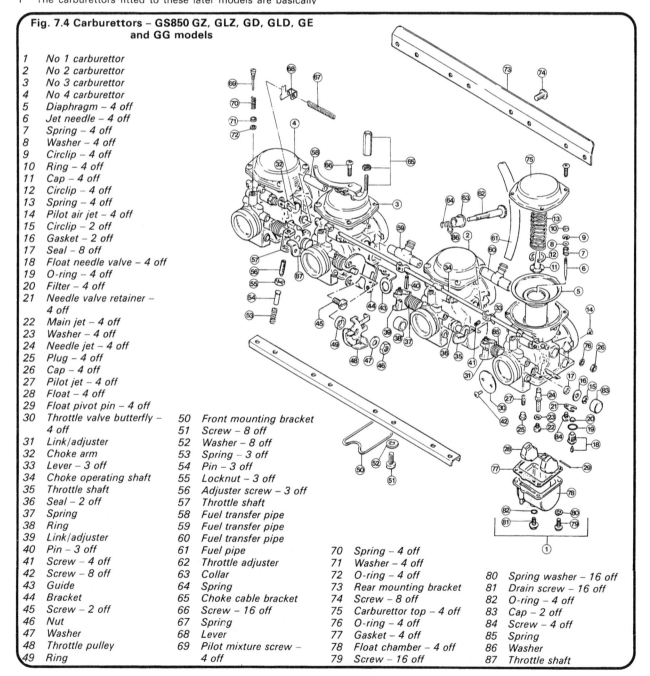

Fig. 7.4 Carburettors – GS850 GZ, GLZ, GD, GLD, GE and GG models

1   No 1 carburettor
2   No 2 carburettor
3   No 3 carburettor
4   No 4 carburettor
5   Diaphragm – 4 off
6   Jet needle – 4 off
7   Spring – 4 off
8   Washer – 4 off
9   Circlip – 4 off
10 Ring – 4 off
11 Cap – 4 off
12 Circlip – 4 off
13 Spring – 4 off
14 Pilot air jet – 4 off
15 Circlip – 2 off
16 Gasket – 2 off
17 Seal – 8 off
18 Float needle valve – 4 off
19 O-ring – 4 off
20 Filter – 4 off
21 Needle valve retainer – 4 off
22 Main jet – 4 off
23 Washer – 4 off
24 Needle jet – 4 off
25 Plug – 4 off
26 Cap – 4 off
27 Pilot jet – 4 off
28 Float – 4 off
29 Float pivot pin – 4 off
30 Throttle valve butterfly – 4 off
31 Link/adjuster
32 Choke arm
33 Lever – 3 off
34 Choke operating shaft
35 Throttle shaft
36 Seal – 2 off
37 Spring
38 Ring
39 Link/adjuster
40 Pin – 3 off
41 Screw – 4 off
42 Screw – 8 off
43 Guide
44 Bracket
45 Screw – 2 off
46 Nut
47 Washer
48 Throttle pulley
49 Ring

50 Front mounting bracket
51 Screw – 8 off
52 Washer – 8 off
53 Spring – 3 off
54 Pin – 3 off
55 Locknut – 3 off
56 Adjuster screw – 3 off
57 Throttle shaft
58 Fuel transfer pipe
59 Fuel transfer pipe
60 Fuel transfer pipe
61 Fuel pipe
62 Throttle adjuster
63 Collar
64 Spring
65 Choke cable bracket
66 Screw – 16 off
67 Spring
68 Lever
69 Pilot mixture screw – 4 off

70 Spring – 4 off
71 Washer – 4 off
72 O-ring – 4 off
73 Rear mounting bracket
74 Screw – 8 off
75 Carburettor top – 4 off
76 O-ring – 4 off
77 Gasket – 4 off
78 Float chamber – 4 off
79 Screw – 16 off

80 Spring washer – 16 off
81 Drain screw – 16 off
82 O-ring – 4 off
83 Cap – 2 off
84 Screw – 4 off
85 Spring
86 Washer
87 Throttle shaft

7 Air filter: cleaning – all models except GS850 GX

1 All later models (except the GS850 GX) use slightly modified air filter side covers which require an altered procedure to remove and refit the element.
2 It should only be necessary to remove the right-hand side panel and filter side cover (the latter secured by a single screw on its bottom edge) to gain access to the element. Lift the element slightly and withdraw from its casing. If, however, the element assembly proves difficult to remove, the left-hand side cover components must be removed so that the element assembly can be pushed out.
3 Once removed the element can be detached from its carrier, cleaned and re-oiled as described in Chapter 2.13.
4 On refitting, ensure that the locating tabs on the carrier left-hand end slide fully into their anchorages so that the element assembly is securely clamped and unfiltered air cannot leak past.

8 Exhaust system: removal and refitting

All models – general
1 On all models covered in this manual, note that by far the easiest method of removal consists simply of lowering the system as a single unit onto a protective layer of rag or newspaper. For all service operations which require the removal of all or part of the exhaust system, this is much quicker than attempting to dismantle the system when it is in place on the machine.
2 Always start by unscrewing the exhaust pipe to cylinder head bolts. Note that these are prone to corrosion and often attempts at removal result in damage to the cylinder head, or sheared bolts; apply penetrating fluid prior to attempting removal. If a bolt proves particularly stubborn the application of heat by an expert will usually release it. With the bolts removed, dismantle the balance box rubber mountings (where fitted), and lower the side stand. Place a thick protective layer of rag or similar on the ground under the machine, then carefully unscrew the silencer mounting bolts or nuts and lower the system to the floor.
3 If required, the system can be manoeuvred from under the machine and is then ready to be dismantled and cleaned. Again, the clamp bolts will almost certainly require the use of penetrating fluid if they are to be unscrewed. Always renew all seals and gaskets, but particularly those in the cylinder head exhaust ports.
4 On reassembly, thoroughly clean all threads until the fasteners can be easily screwed together by hand, then apply a good layer of heatproof anti-corrosion compound such as Copaslip to all nuts and bolts to prevent the formation of corrosion in the future. Use a smear of grease to stick the new exhaust gaskets in place. If the system has been dismantled, assemble it loosely under the machine and around the side stand; ensure that all clamps are refitted the correct way round before the pipes are assembled. Note that separate pipes (where applicable) are usually stamped with the cylinder number to ensure correct assembly, also that the pipe clamps have locating tabs to ensure that they are positioned correctly.
5 Enlisting the aid of an assistant to lift the silencer rear ends, lift the system into place and loosely refit the cylinder head bolts and silencer mountings. Check that all components are seated correctly and without strain; if adjustment is necessary, gently use a rubber mallet (or a wooden mallet tapping on to rag wrapped around the pipe) to tap components into place. *Excessive force should not be required.* Unless components have been renewed or perfectly cleaned, their original locations can usually be identified from the marks left in the dirt or corrosion.
6 Tighten the cylinder head bolts to the specified torque setting, then tighten the pipe clamps and balance box mountings (where fitted) securely. Finally tighten the silencer mounting bolts or nuts and check that all components are secure, particularly the small covers over the pipe/silencer joints, which can cause an annoying rattle if loose.

GS850 GX, GLX, GZ, GLZ, GLD models
7 These models all use basically the same type of system as described in Chapter 1 for the GS850 GT and GLT models. That is two outer units, comprising an integral exhaust pipe and silencer, are joined by a balance box which is rubber-mounted at three points under the engine/gearbox unit. The two middle cylinders are served by separate pipes running into the balance box.

GS850 GD, GE, GG models
8 The system fitted to these later models uses again two outer units, as described above. The middle cylinder exhaust pipes, however, are now integral with the balance box; the balance box rubber mountings are deleted.

9 Ignition system: general description and testing – GS850 GZ, GD, GLD, GE and GG models

General description
1 These later models are fitted with an all-electronic ignition system in which the last remaining mechanical component, the ATU, is replaced by a signal generator rotor; the advance curve is now controlled more precisely by the signals generated by the pick-up/pulser coils acting on circuits within the ignitor.
2 The signal generator assembly is similar to that described for the GS850 GT and GLT models in Chapter 3. Note however that the rotor can be removed and refitted without disturbing the generator backplate, also that there is little or no provision for altering the position of the backplate and thus the ignition timing. The system no longer contains any mechanical components, which are subject to wear, and regular checks of the ignition timing are less necessary than for the previous models.
3 The ignitor is mounted behind the left-hand side panel. On GS850 GZ models its connector is the rearmost (darker-coloured) of the three next to the fuse box and the unit itself is mounted under the battery/electrical component carrier panel. On GS850 GD, GLD, GE and GG models the connector plugs directly into the unit (except for the GS850 GLD, on which the connector is next to the unit), which is mounted between the regulator/rectifier and the starter motor relay. In all cases the ignitor is a sealed black plastic unit which is retained by two screws.

Testing the ignition system
4 As electronic components are generally reliable, problems with the ignition system will usually be confined to faulty spark plugs or suppressor caps, faulty switches or damaged wiring. Tracing a fault can be considerably eased if the symptoms are considered carefully. For example if the fault occurs only in wet weather, use a water-dispersant spray to seal the suppressor caps, HT leads, switches and all connections until the fault is eliminated; careful sealing should then achieve a cure.
5 If a fault is confined only to cylinders 1 and 4, or to cylinders 2 and 3, then it is most likely to be traced to the pick-up coil or HT coil (or their wiring) for that pair of cylinders. Similarly total failure, if not caused by a failure of the power supply from the battery, is mosy likely to be due to a fault in the ignition or kill switches or in the ignitor itself.
6 Given below is a simple test sequence which can be applied with the minimum of special tools (although a good multimeter would be very useful) and should quickly eliminate any fault.

a) Check that the ignition and engine kill switches are in the 'On' and 'Run' positions respectively
b) Check that all other electrical systems are working to ensure that the battery is in good condition
c) Check that the ignition circuit fuse and main fuse are sound

d) Check for a strong blue spark at all four spark plugs, ensuring that the plug metal bodies are securely earthed to the cylinder head while the test is made. If no spark, or a weak spark, is found first substitute brand new spark plugs; if this fails to cure the fault, check the suppressor caps by unscrewing them and holding the HT lead bared end approximately 6 mm ($\frac{1}{4}$ in) away from the cylinder head. A strong blue spark should jump the gap when the engine is turned over. **Note:** test one lead at a time and ensure that the remaining three are securely earthed at all times via their suppressor caps and spark plugs, to prevent any risk of damage to the system components. Also, be very careful to hold the lead as described; the spark may well find it easier to travel to earth by way of your hand, thus producing a most unpleasant shock!

e) Check the power supply to the HT coils by disconnecting their Orange/white wires; with the ignition switched on full battery voltage should be available, measured between the wire terminals and a good earth point. While this test is best conducted using a multimeter set to the appropriate scale, a 12 volt bulb can be used as a test lamp to achieve equally good results. If full voltage is not available check back through the wiring via the starter button, engine kill switch and fuses to the ignition switch, until the fault is located and rectified.

f) Check the wiring between the signal generator pick-up coils, ignitor and HT coils, looking for obvious signs of damage. If a meter or battery and bulb test circuit is available the continuity of the wires themselves can be checked to ensure that there are no breaks or short circuits.

g) If all other possibilities have been checked and the fault persists, it must be in one or more of the system components themselves. These should be checked in the following order:
   1) Signal generator unit – see paragraph 7
   2) Ignition HT coils – see Chapter 3.3 for full test. Specifications Section of Chapers 3 and 7 for simple resistance test
   3) Ignitor unit – see paragraph 8

Signal generator – testing
7 This unit is tested using the same basic procedure given in Chapter 3.12, but noting the revised resistances and test connections given in the Specifications Section of this Chapter. Note also the location of the signal generator/ignitor connector plug, as described in paragraph 3 of this Section.

Ignitor – testing
8 This unit is tested by following the basic procedure given in Chapter 3.11, ie using a multimeter set to the x 1 ohm resistance scale as a substitute for the signal generator. Note that the HT coils should be known to be sound before carrying out this check. If no spark is produced, then the ignitor is proven faulty by elimination and must be renewed. Proceed as follows, noting the different location of the signal generator/ignitor connector plug, as described in paragraph 3 of this Section.
9 Disconnect the signal generator connector plug, then set the spark plugs, ignition switch and meter as described in Chapter 3. Connect the meter probes to the unit's terminals as follows:
10 With the meter positive (+) probe touched to the unit Blue wire terminal, and its negative (−) probe touched to the unit Yellow wire terminal, then a spark should occur at No 1 cylinder spark plug as soon as both probes are removed simultaneously.
11 With the meter positive (+) probe touched to the unit Green wire terminal, and its negative (−) probe touched to the unit Black wire terminal, then a spark should occur at No 2 cylinder spark plug as soon as both probes are removed simultaneously.
12 If no sparks are produced, and the ignition HT coils, suppressor caps and spark plugs are known to be in good condition, then the ignitor is proved faulty by elimination.

10 Front forks: modifications

GS850 GZ (UK), GE, GG models
1 Apart from the separate bush described below, the forks fitted to these models are similar in design and layout to those described for the UK models in Chapter 4. Note the different fork oil quantity and spring free length given in the Specifications Section of this Chapter, also that the spring preload adjuster has four positions instead of the three on the GT and GX models.
2 These models are fitted with forks in which the lower leg bears on a replaceable bush rather than on the stanchion itself. To dismantle these forks proceed as described in Chapter 4.3, paragraphs 1 and 2. Difficulty may however be encountered in removing the stanchion from the lower leg so great care is needed.
3 The manufacturer states that the stanchion, complete with bottom bush, can be pulled through the oil seal, once the damper rod Allen bolt has been removed. Owners should note, however, that this will certainly damage the seal's delicate lips, thus the seal must be renewed whenever it is disturbed in this way. On refitting, the stanchion and bush should be inserted *first* into the lower leg, then the seal should be fitted over the stanchion upper end and slid down until it can be tapped evenly into its housing in the lower leg.
4 **Note:** The usual arrangement with 'bushed' forks is that a top bush is pressed into the top of the lower leg followed by a shaped backing washer and then the oil seal. If a top bush is encountered, the following alternative dismantling procedure should be noted:
5 Remove the damper rod Allen bolt, the dust excluder and the oil seal retaining circlip as described in Chapter 4.3, then clamp the lower leg by the spindle lug in a soft-jawed vice. Push the stanchion fully into the lower leg then pull it sharply out. Repeat as necessary, using the slide hammer action of the bottom bush against the top bush, until the oil seal and backing washer (where fitted) are ejected and the stanchion can be withdrawn. Note that the top bush and backing washer are not listed separately and cannot therefore be renewed if damaged; the advice of a good Suzuki Service Agent must be obtained if this situation is encountered.
6 Once the stanchion has been removed, the bottom bush can be withdrawn by inserting a slim screwdriver blade into its split seam and prising the ends apart just far enough to permit the bush to be slid off the stanchion lower end. The bushes should be renewed together whenever signs of excessive wear or damage can be seen on their bearing surfaces.
7 Reassembly is the reverse of the dismantling procedure. Apply liberal quantities of fork oil to aid assembly and note that for both types the oil seal must be fitted after the stanchion and bush have been installed.

GS850 GZ (US), GLX, GLZ, GLD models
8 These models all use front forks which differ from the GT/GX or GLT unit (as appropriate) only in the fitting of a separate bottom bush. Refer to paragraphs 2 to 6 above for instructions on removal and refitting of the stanchion and bush. On GS850 GLX, GLZ and GLD models, note that the fork springs must be refitted with their tapered ends downwards and that the fork legs are refitted to the machine so that the groove in the stanchion upper end aligns with the top surface of the fork top yoke. Note also the differing fork oil quantities and spring free lengths given in the Specifications Section of this Chapter.

GS850 GD model
9 This model uses front forks which are similar to the GT units described in Chapter 4, but are fitted with separate bottom bushes and have a hose linking the air chamber in each leg via holes in the stanchions. The separate top plug/air caps of the earlier models are replaced by a large threaded plug at the top of each stanchion, and a valve on the hose left-hand end,

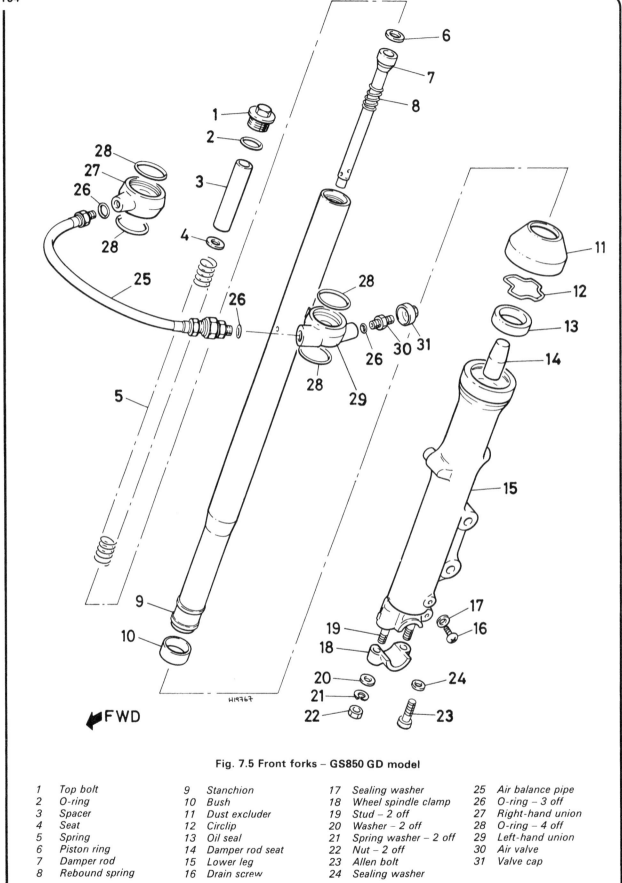

H19767

←FWD

Fig. 7.5 Front forks – GS850 GD model

1	Top bolt	9	Stanchion	17	Sealing washer	25	Air balance pipe
2	O-ring	10	Bush	18	Wheel spindle clamp	26	O-ring – 3 off
3	Spacer	11	Dust excluder	19	Stud – 2 off	27	Right-hand union
4	Seat	12	Circlip	20	Washer – 2 off	28	O-ring – 4 off
5	Spring	13	Oil seal	21	Spring washer – 2 off	29	Left-hand union
6	Piston ring	14	Damper rod seat	22	Nut – 2 off	30	Air valve
7	Damper rod	15	Lower leg	23	Allen bolt	31	Valve cap
8	Rebound spring	16	Drain screw	24	Sealing washer		

immediately above the bottom yoke, permits adjustment of the air pressure.

10 Using the accompanying illustration for details, refer to either Chapter 4.3 or to paragraphs 2 to 6 of this Section, as necessary, for information on dismantling and rebuilding this type of fork.

11 On refitting the fork legs to the machine, apply a thin smear of grease to the stanchion upper end and take care not to dislodge or damage any of the O-rings sealing the hose unions. To prevent the loss of air pressure these O-rings should be renewed if there is any doubt about their condition.

11 Steering head: modifications

1 With reference to Fig. 4.3, note that item 9 is replaced by a bolt on all GS850 GZ, GLZ, GD, GLD, GE and GG models.
2 Also with reference to Fig. 4.3, note that GS850 GLX models are shown as having a washer fitted between the steering head bearings (items 3 and 24); if the bearings are renewed or disturbed at any time ensure that the washer is not omitted.

12 Rear suspension units: modifications

Note that on all later models (except the GS850 GX) the right-hand suspension unit bottom mounting is now by a special bolt rather than the cap nut formerly fitted.

13 Footrests: general – all models except GS850 GX

1 Note that on all later models the footrests are rubber-mounted on separate mounting plates which are bolted directly to the frame. If the footrests are dismantled note the arrangement of mounting rubbers and washers so that they can be correctly installed on reassembly; it is essential that the footrests are flexibly-mounted when the bolts are fully tightened if the rider is to be protected from vibration.
2 Note also that on most models the pillion footrests and silencer mountings are combined.

14 Front wheel: modification – GS850 GZ, GD, GE and GG models

With reference to Fig. 5.1 note that the wheel right-hand spacer (item 8) is a plain tubular type on these models. Ensure that the front forks are correctly aligned on the spacer by bouncing the suspension up and down several times before tightening the spindle clamp nuts to the specified torque setting.

15 Rear wheel: removal, refitting and modifications

Removal and refitting – except GS850 GX model
1 On all later models except the GS850 GX, the rear wheel removal procedure is slightly altered. Proceed as follows:

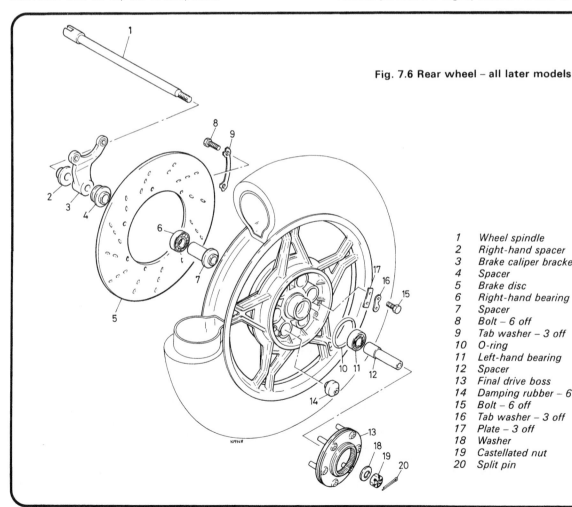

Fig. 7.6 Rear wheel – all later models

1 Wheel spindle
2 Right-hand spacer
3 Brake caliper bracket
4 Spacer
5 Brake disc
6 Right-hand bearing
7 Spacer
8 Bolt – 6 off
9 Tab washer – 3 off
10 O-ring
11 Left-hand bearing
12 Spacer
13 Final drive boss
14 Damping rubber – 6 off
15 Bolt – 6 off
16 Tab washer – 3 off
17 Plate – 3 off
18 Washer
19 Castellated nut
20 Split pin

2   Place the machine on its centre stand on level ground, then remove the left-hand rear suspension unit. Slacken the right-hand unit's top mounting nut and unscrew its bottom mounting bolt completely.

3   Pull the right-hand rear suspension unit clear of its bottom mounting, then lift the swinging arm until the wire strap from the toolkit can be slipped over both mounting lugs of the left-hand unit. The swinging arm should now be supported securely so that the rear wheel spindle is raised clear of the silencers.

4   Straighten and remove the split pin from the wheel spindle end and unscrew the spindle nut. Slacken the pinch bolt in the swinging arm right-hand fork end. Straighten and remove its split pin, then remove the rear brake caliper torque arm nut and bolt.

5   On GS850 GZ, GD, GE and GG models the caliper can be withdrawn complete with its bracket, as soon as the wheel spindle is removed, and can be then be hung from the frame seat tubes using a length of wire or string if there are no convenient projections; take care not to twist or damage the hose. On GS850 GLX, GLZ and GLD models if the pinch bolt/brake hose clamp bolt is first removed completely, a similar procedure can be used. The alternative is to remove the two caliper/bracket bolts and to hang the caliper over the silencer; be careful not to twist or damage the hose, ensure that the silencer has fully cooled down before placing the caliper or hose near it, and use a thick piece of rag to protect the finish of all components.

6   On all models, as soon as the caliper is removed from the disc be careful not to apply the brake pedal or the pads and pistons may be dislodged, causing seal damage and fluid loss. The risk of this can be minimised by wedging a clean spacer (of the same thickness as the disc) between the pads whenever the caliper is removed.

7   When the caliper is separated from the torque arm (and bracket) withdraw the wheel spindle noting the two spacers, one on each side of the caliper bracket. Pull the wheel to the right, off the splined coupling and manoeuvre it away from the machine.

8   On reassembly thoroughly clean and grease the wheel spindle and apply a thin smear of grease to the coupling splines. Check that the threads of all fasteners are clean and apply anti-corrosion compound such as Copaslip to any that show signs of corrosion.

9   Refitting is the reverse of the removal procedure. Ensure that the spacers are refitted correctly on each side of the caliper bracket, and do not forget to replace the caliper on its bracket (where applicable) and torque arm before tightening the wheel spindle nut. Ensure that all fasteners are tightened securely to the specified torque settings, where possible, and that new split pins are correctly fitted to secure the spindle and torque arm nuts.

10 Apply the brake pedal several times until the pads are back in firm contact with the disc and normal pressure is restored. Before taking the machine out on the road check that the rear brake and suspension work properly and that the wheel is free to rotate with the minimum of drag from the rear brake or final drive.

Modifications

11 With reference to the accompanying illustration, note that the wheel spindle on all later models (except the GS850 GX) is fitted from right to left, and that the final drive coupling on all models is slightly altered from that shown in Chapter 5.

16 Brakes: modifications

1   Although the front brake master cylinder fitted to GS850 GZ, GLZ, GD, GLD, GE and GG models is different in appearance from those described in Chapter 5, there are no significant differences in working procedure.

2   With reference to Chapter 5.6, paragraph 4, the master cylinder dust boot is no longer secured by a stopper – it can be

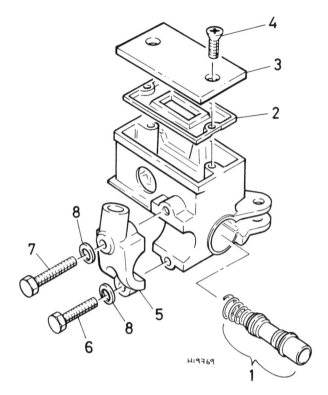

Fig. 7.7 Front brake master cylinder – GS850 GZ, GD, GE and GG models (GLZ and GLD models similar)

1	Piston/seal assembly	5	Clamp
2	Diaphragm	6	Bolt
3	Filler cap	7	Bolt
4	Screw	8	Washer – 2 off

pulled carefully away as soon as the brake lever has been removed.

3   With reference to Fig. 5.8 (items 11 and 12) and to Chapter 5.12 paragraph 3, the rear brake caliper fitted to GS850 GZ, GD, GE and GG models is fitted with only one bleed nipple, on the inboard side. This simplifies the bleeding procedure.

4   With reference to Fig. 5.9 (item 21) the rear brake hose clamp and bolt are deleted from GS850 GZ, GD, GE and GG models.

17 Tubeless tyres: removal and refitting – all US models except GS850 GX

1   It is strongly recommended that should a repair to a tubeless tyre be necessary, the wheel is removed from the machine and taken to a tyre fitting specialist who is willing to do the job or taken to a Suzuki Service Agent. This is because the force required to break the seal between the wheel rim and tyre bead is considerable and considered to be beyond the capabilities of an individual working with normal tyre removing tools. Any abortive attempt to break the rim to bead seal may also cause damage to the wheel rim, resulting in an expensive wheel replacement. If, however, a suitable bead releasing tool is available, and experience has already been gained in its use, tyre removal and refitting can be accomplished as follows.

2   Remove the wheel from the machine as described in Chapter 5.3 (front) or Section 16 (rear), as appropriate. Deflate the tyre by removing the valve insert and when it is fully deflated, push the bead of the tyre away from the wheel rim on both sides so that the bead enters the centre well of the rim. As noted, this operation will almost certainly require the use of a bead releasing tool.

3   Insert a tyre lever close to the valve and lever the edge of the tyre over the outside of the wheel rim. Very little force should be necessary: if resistance is encountered it is probably due to the fact that the tyre beads have not entered the well of the wheel rim all the way round the tyre. Should the initial problem persist, lubrication of the tyre bead and the inside edge and lip of the rim will facilitate removal. Use a recommended lubricant, a diluted solution of washing-up liquid or french chalk. Lubrication is usually recommended as an aid to tyre fitting but its use is equally desirable during removal. The risk of lever damage to wheel rims can be minimised by the use of proprietary plastic rim protectors placed over the rim flange at the point where the tyre levers are inserted. Suitable rim protectors may be fabricated very easily from short lengths (4 - 6 inches) of thick-walled nylon petrol pipe which have been split down one side using a sharp knife. The use of rim protectors should be adopted whenever levers are used, and, therefore, when the risk of damage is likely.

4   Once the tyre has been edged over the wheel rim, it is easy to work around the wheel rim so that the tyre is completely freed on one side.

5   Working from the other side of the wheel, ease the other edge of the tyre over the outside of the wheel rim, which is furthest away. Continue to work around the rim until the tyre is freed completely from the rim.

6   Refer to the following Section for details relating to puncture repair and the renewal of tyres. See also the remarks relating to the tyre valves in Section 19.

7   Refitting of the tyre is virtually a reversal of the removal procedure. If the tyre has a balance mark (usually a spot of coloured paint) indicating its lightest point, as on the tyres fitted as original equipment, this must be positioned alongside the valve. Similarly any arrow indicating direction of rotation must be fitted pointing the right way. If only one arrow is found this must be fitted pointing in the direction of rotation on **rear** wheels, but **opposite** to the direction of rotation, to accept braking loads, on **front** wheels. Where two arrows (marked 'Front Wheel' and 'Rear Wheel') are provided, fit the tyre as indicated by them.

8   Starting at the point furthest from the valve, push the tyre bead over the edge of the wheel rim until it is located in the central well. Continue to work around the tyre in this fashion until the whole of one side of the tyre is on the rim. It may be necessary to use a tyre lever during the final stages. Here again, the use of a lubricant will aid fitting. It is recommended strongly that when refitting the tyre only a recommended lubricant is used because such lubricants also have sealing properties. Do not be over generous in the application of lubricant or tyre creep may occur.

9   Fitting the upper bead is similar to fitting the lower bead. Start by pushing the bead over the rim and into the well at a point diametrically opposite the tyre valve. Continue working round the tyre, each side of the starting point, ensuring that the bead opposite the working area is always in the well. Apply lubricant as necessary. Avoid using tyre levers unless absolutely essential, to help reduce damage to the soft wheel rim. The use of the levers should be required only when the final portion of bead is to be pushed over the rim.

10 Lubricate the tyre beads again prior to inflating the tyre, and check that the wheel rim is evenly positioned in relation to the tyre beads. Inflation of the tyre may well prove impossible without the use of a high pressure air hose. The tyre will retain air completely only when the beads are firmly against the rim edges at all points and it may be found when using a foot pump that air escapes at the same rate as it is pumped in. This problem may also be encountered when using an air hose on new tyres which have been compressed in storage and by virtue of their profile hold the beads away from the rim edges. To overcome this difficulty, a tourniquet may be placed around the circumference of the tyre, over the central area of the tread. The compression of the tread in this area will cause the beads to be pushed outwards in the desired direction. The type of tourniquet

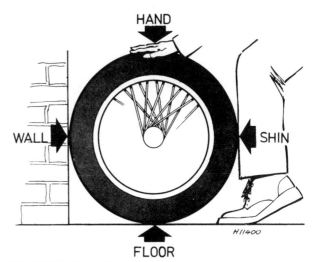

Fig. 7.8 Method of seating the beads on tubeless tyres

most widely used consists of a length of hose closed at both ends with a suitable clamp fitted to enable both ends to be connected. An ordinary tyre valve is fitted at one end of the tube so that after the hose has been secured around the tyre it may be inflated, giving a constricting effect. Another possible method of seating beads to obtain initial inflation is to press the tyre into the angle between a wall and the floor. With the airline attached to the valve additional pressure is then applied to the tyre by the hand and shin, as shown in the accompanying illustration. The application of pressure at four points around the tyre's circumference whilst simultaneously applying the airhose will often effect an initial seal between the tyre beads and wheel rim, thus allowing inflation to occur.

11 Having successfully accomplished inflation, increase the pressure to 40 psi and check that the tyre is evenly disposed on the wheel rim. This may be judged by checking that the thin positioning line found on each tyre wall is equidistant from the rim around the total circumference on the tyre. If this is not the case, deflate the tyre, apply additional lubrication and reinflate. Minor adjustments to the tyre position may be made by bouncing the wheel on the ground.

12 Always run the tyre at the recommended pressures and never under or over-inflate. The correct pressures for original equipment tyres are given in the Specifications Section of this Chapter or Chapter 5; check with the tyre manufacturers (if non-standard tyres are fitted) and use their recommended pressures, if different. Remember to check the wheel balance as described in Chapter 5, Section 20.

18 Tubeless tyres: puncture repair and tyre renewal – all US models except GS850 GX

1   If a puncture occurs, the tyre should be removed for inspection for damage before any attempt is made at remedial action. The temporary repair of a punctured tyre by inserting a plug from the outside should **not** be attempted. Although this type of temporary repair is used widely on cars, the manufacturers strongly recommend that no such repair is carried out on a motorcycle tyre. Not only does the tyre have a thinner carcass, which does not give sufficient support to the plug, but the consequences of a sudden deflation are often sufficiently serious that the risk of such an occurrence should be avoided at all costs.

2   The tyres should be inspected both inside and out for damage to the carcass. Unfortunately the inner lining of the tyre – which takes the place of the inner tube – may easily obscure any damage and some experience is required in making a

correct assessment of the tyre condition. **Note:** *If the puncture is in the sidewalls, or if it is greater than 6 mm (0.24 in) in diameter, do not attempt a repair – the tyre must be renewed.*

3   If the puncture is small enough, and in the main tread area, there are two main types of tyre repair which are considered safe for adoption in repairing tubeless motorcycle tyres. The first type of repair consists of inserting a mushroom-headed plug into the hole from the inside of the tyre. The hole is prepared for insertion of the plug by reaming and the application of an adhesive. The second repair is carried out by buffing the inner lining in the damaged area and applying a cold or vulcanised patch. Because both inspection and repair, if they are to be carried out safely, require experience in this type of work, it is recommended that the tyre be placed in the hands of a repairer with the necessary skills, rather than repaired in the home workshop.

4   A repaired tyre must not be used at speeds of more than 50 mph (80 km/h) for 24 hours after the repair; the repair may fail, allowing the tyre to deflate again. It should also **never** be used at speeds of more than 80 mph (130 km/h) especially when the machine is heavily loaded.

5   Owners stranded with no means of repairing a punctured tubeless tyre should note that the commonly-used trick of fitting a tube is not recommended by Suzuki or by the tyre manufacturers. If such a course of action is adopted as an emergency repair, against this advice, note that the cause of the puncture must be removed first, the tube valve stem must be supported in the rim by the fitting of a suitable spacer and the machine must be ridden as slowly as possible so as not to cause overheating of the tube. Certain makes of tubeless tyre use large internal supporting ribs which will rapidly destroy an inner tube due to chafing.

19 Tubeless tyre valves: examination and renewal – all US models except GS850 GX

1   The valve core is of the same type as that used with tubed tyres, and screws into the valve body. The core can be removed with a small slotted tool which is normally incorporated in plunger tyre pressure gauges. Some valve dust caps incorporate a projection for removing valve cores.

2   If the core is proved to be sound, make a similar check to ensure that air is not leaking from the valve body or from its sealing face. If this is found to be the case, the valve must be renewed, although it is worth checking that the early threaded-type valve's retaining nut (where fitted) is securely fastened.

3   To renew a valve, remove the wheel from the machine and take off the tyre (Section 17). On the early threaded-type valve unscrew the locking nut to release the valve and ensure that it is securely fastened on reassembly. The later car-type rubber-bodied valve must be removed by cutting off its inner retaining shoulder and pulling the remains out of the rim. To fit a new valve, lubricate it thoroughly and push it through the rim from the outside towards the centre (having first removed its cap). It must then be drawn into place by screwing the correct tool on to its threaded end to provide purchase (the tool should be available at any tyre-fitting establishment). If great care is taken to avoid crushing the valve end it is possible to screw a suitably-sized nut on to the valve threads and to grip this with a large pair of pliers. Check that the rubber locating shoulders lock securely into place on each side of the rim. Refit the tyre and wheel.

4   In all other respects, these valves are similar to the tubed type described in Chapter 5, Sections 18 and 19.

20 Electrical system: modifications

General
1   While the electrical system has remained basically the same, many minor changes have been made which means that

very few components are interchangeable between models. Note that this Chapter only mentions those components which require different working procedures.

2   Note particularly that the regulator/rectifier unit has been modified several times; always ensure that the correct item is obtained, should renewal ever prove necessary, so that it is correctly matched to the generator output.

3   GS850 GE and GG models are fitted with a different starter motor relay which is mounted at the rear end of the electrical components panel. Test procedures remain as described in Chapter 6.

Horn – GS850 GLX, GLZ, GLD models
4   Note that these models are fitted with only one horn, mounted vertically on the front of the frame front downtubes.

Horn – GS850 GZ, GD, GE, GG models
5   While these models are all fitted with twin horns, note that they are now rubber mounted on separate brackets, to the front of the frame front downtubes.

Fuses – GS850 GZ, GD, GLD, GE, GG models
6   Later models are fitted with spade-type fuses in place of the conventional tubular type used on previous models. Whilst this does not have any material effect on the function of the fuse, it does mean that only the correct replacement fuse can be used in the event of a failure. It follows that replacement fuses of the correct type and rating should be carried at all times.

Flasher relay unit
7   Note that on most later models, the location of this relay has been moved to behind the right-hand side panel, next to the self-cancelling unit and rear brake master cylinder.

21 Starter motor: overhaul GS850 GZ, GD, GLD, GE and GG models

1   All these later models are fitted with starter motors of the four-brush type. Removal and refitting are as described in Chapter 6.9. Overhaul procedures are given below; refer to the accompanying illustration for details.

2   First mark the left and right-hand ends of the motor body and the end covers so that all can be refitted in their original locations.

3   Remove the two long retaining screws and withdraw the motor left-hand cover noting the sealing O-rings around its locating boss and its mating surface; check carefully the number and position of any shims fitted.

4   Carefully withdraw the motor right-hand cover; again, note the sealing O-ring and the number and position of any shims fitted.

5   Carefully ease the springs out of the brush holders, then withdraw the brushes themselves. Remove the brush holders plate with the two negative (–) brushes; if the plate is to be refitted, carefully clean the brush holders so that each brush is completely free to slide in its holder.

6   Push the armature out of the motor body. Unscrew the terminal retaining nut, withdraw the metal and insulating washers and the O-ring, then remove the field coil brush assembly and plastic insulator.

7   Measure the length of each brush; they are worn out if reduced to a length of 6 mm (0.24 in) or less. Note that one pair of brushes is soldered to the motor terminal and must be renewed with the terminal, while the other pair is crimped to the brush holder plate; this also must be renewed as a single assembly.

8   Check that the springs exert firm pressure on the brushes, that the brushes are not chipped or damaged and that they bear fully on the commutator; check also that each brush is free to slide easily in its holder. The springs are available separately.

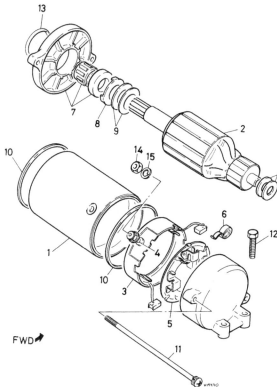

FWD

Fig. 7.9 Starter motor – GS850 GZ, GD, GLD, GE and GG models

1	Motor body	9	Shims
2	Armature	10	Sealing ring – 2 off
3	Brush assembly	11	Screw – 2 off
4	O-ring	12	Bolt – 2 off
5	Brush holder plate	13	O-ring
6	Brush spring – 4 off	14	Nut
7	Drive housing	15	Washer
8	Thrust support		

9 Clean the commutator segments with a rag soaked in methylated spirits and inspect each one for scoring or discoloration. If any pair of segments is discoloured, a shorted armature winding is indicated. Clean out the grooves between the commutator segments, using a hacksaw blade ground to the correct width. Each segment should be straight sided; if the depth of undercut is less than 0.2 mm (0.008 in) the armature should preferably be renewed. It is possible to re-cut the grooves using the modified hacksaw blade mentioned above, but this requires care and patience. Do not cut into the segment material, or leave the groove anything other than square sided.
10 The cleaning of the commutator segments with abrasive paper is not recommended due to the risk of particles becoming embedded in the soft segments. It is suggested, therefore, that an ink eraser be used to burnish the segments and remove any surface oxide deposits before installing the brushes.
11 Using a multimeter set on the resistance scale, check for continuity between each segment and its neighbours, noting that anything other than a very low resistance indicates a partially or completely open circuit. Next check the armature insulation by checking for continuity between the armature core and each segment. Anything other than infinite resistance indicates an internal failure. If any fault is found, the armature must be renewed.
12 Check each field coil brush by testing for continuity between it and the terminal; no resistance should be encountered. Check also that the terminal is completely insulated from the motor body. Finally, check that continuity (ie little or no resistance) exists between the field coils, then that each coil is completely insulated from the motor body. If

continuity is found between the field coils and the body there is a breakdown in insulation which means that the complete assembly must be renewed.
13 If oil is found in the starter motor, the seal pressed into the left-hand cover is faulty and must be renewed. Check the bearing at each end of the armature by reassembling the motor and feeling for free play. Spin the left-hand bearing and check for signs of roughness, wear, or other damage. The bearing must be renewed if at all worn, but note that neither this nor the seal are available separately; the apparent solution is to purchase a new cover assembly. To avoid unnecessary expense, an automotive parts supplier or specialist bearing supplier may be able to find suitable replacements. Ensure that all relevant dimensions and seal or bearing markings are noted so that the correct items can be selected; if necessary take the motor assembly to provide a pattern.

22 Starter clutch: modification – all models except GS850 GX

With reference to Fig. 6.5 items 1, 2 and 3, note that on all later models except the GS850 GX the idler gear has been widened and the shaft lengthened, also that the two thrust washers are deleted.

23 Side stand warning lamp: general – GS850 GZ, GLZ, GD, GLD, GE and GG models

1 These models are fitted with a warning lamp which lights whenever the ignition is switched on and the side stand is down. The circuit consists simply of a plunger-type switch mounted on the side stand pivot and the lamp in the instrument panel itself.
2 If the lamp ever fails to light, first check the bulb itself. If this proves sound, check that full battery voltage is available at the bulb holder; if not; check back along the Orange (or Orange/green) wires to the ignition switch and fuses until the fault is found and rectified.
3 If the feed is correct at the bulb holder check the Green/white wire from the bulb holder to the switch, also the Black/white wire from the switch to earth; if continuity is not found the wires must be checked carefully until the break is found and rectified.
4 In normal circumstances, however, the fault will usually lie in the switch, which is in a very exposed position and will be prone to failure due to the effects of dirt, water and corrosion. To check the switch, disconnect its wires and use a multimeter set to the resistance function or a battery and bulb test circuit to establish that continuity exists when the switch plunger is extended. When the plunger is pressed in the terminals should be isolated.
5 If the switch is faulty it must be renewed; repairs are not possible. To prolong the life of the switch, peel back its protective cover and pack the switch with silicone grease to exclude dirt and water. Refit the cover and spray the switch, especially the plunger, at regular intervals with a water-dispersant lubricant such as WD40.
6 The switch is secured by two screws to the stand pivot. Its position can be adjusted by slackening these screws and moving the switch body until the warning lamp goes out when the side stand is up. Tighten the screws carefully.
7 Note that the side stand warning lamp circuit is connected to the oil pressure warning lamp circuit by a diode. This is a small tubular component which is enclosed in a clear plastic sheath and is to be found in the main wiring loom, in the region of the headlamp. If a fault ever occurs in both circuits, particularly in the feed from the bulb holders to the switches, the diode should be checked using a multimeter. Continuity should exist in one direction only; if continuity is found in both directions, or in neither, the diode is faulty and must be renewed.

24 Instruments: modifications

1   The instrument assemblies fitted to the later models all vary slightly from that shown in Chapter 6. Refer to the relevant accompanying illustration for details.

2   Note particularly that the bulbs fitted to later models are of the capless type. These are a push fit in their holders and care is required on refitting to correctly align their fine wire tails with the bulb holder contacts.

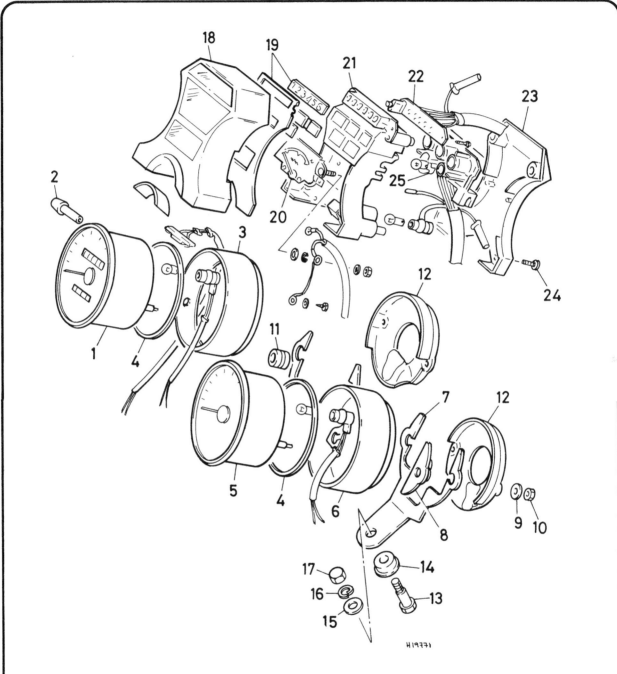

Fig. 7.10 Instrument console – GS850 GLT and GLX models

1	Speedometer	8	Damping rubber – 4 off	15	Washer – 2 off	22	Gear position indicator
2	Tripmeter reset knob	9	Washer – 4 off	16	Spring washer – 2 off	23	Lower cover
3	Speedometer housing	10	Nut – 4 off	17	Nut – 2 off	24	Screw – 4 off
4	Damping ring – 2 off	11	Damping rubber – 2 off	18	Indicator lamp case	25	Bulb holder
5	Tachometer	12	Lower cover – 2 off	19	Case lens		
6	Tachometer housing	13	Bolt – 2 off	20	Fuel gauge		
7	Mounting bracket	14	Damping rubber – 2 off	21	Housing		

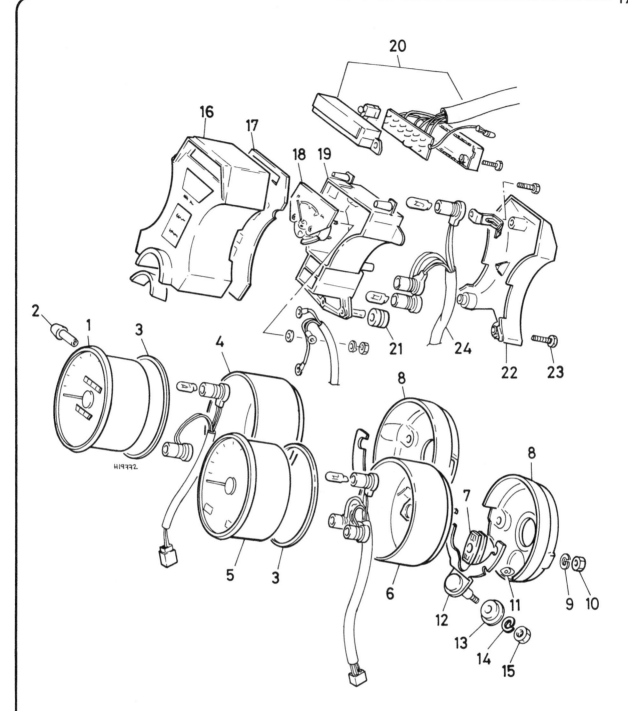

Fig. 7.11 Instrument console – GS850 GLZ and GLD models

1	Speedometer	7	Damping rubber – 4 off	13	Damping rubber – 2 off	19	Housing
2	Tripmeter reset knob	8	Lower cover – 2 off	14	Spring washer – 2 off	20	Gear position indicator
3	Damping ring – 2 off	9	Washer – 2 off	15	Nut – 2 off	21	Damping rubber – 2 off
4	Speedometer housing	10	Nut – 2 off	16	Indicator lamp case	22	Lower cover
5	Tachometer	11	Damping rubber – 2 off	17	Case lens	23	Screw – 3 off
6	Tachometer housing	12	Mounting bracket	18	Fuel gauge	24	Bulb holder

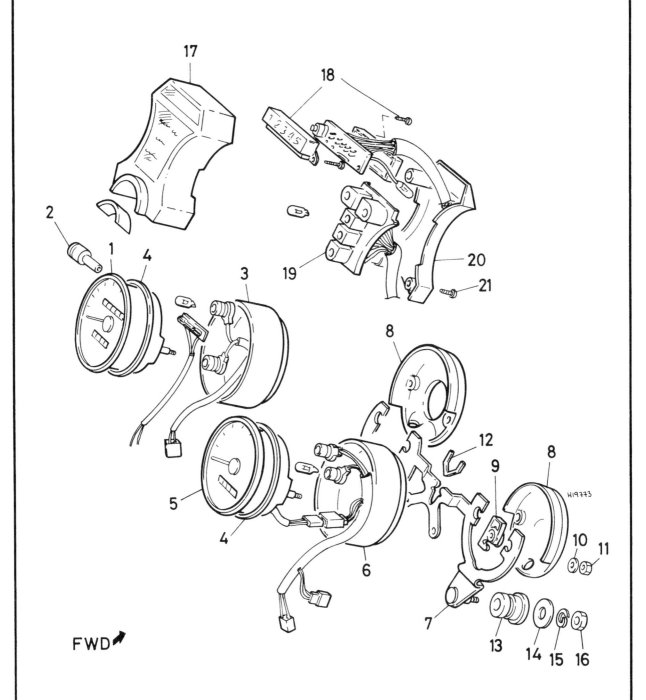

FWD↗

Fig. 7.12 Instrument console – GS850 GZ, GD, GE and GG models

1	Speedometer	7	Mounting bracket	12	Cover	17	Indicator lamp case
2	Tripmeter reset knob	8	Lower cover – 2 off	13	Damping rubber – 2 off	18	Gear position indicator
3	Speedometer housing	9	Damping rubber – 4 off	14	Washer – 2 off	19	Bulb holder
4	Damping ring – 2 off	10	Washer – 4 off	15	Spring washer – 2 off	20	Lower cover
5	Tachometer	11	Nut – 4 off	16	Nut – 2 off	21	Screw – 4 off
6	Tachometer housing						

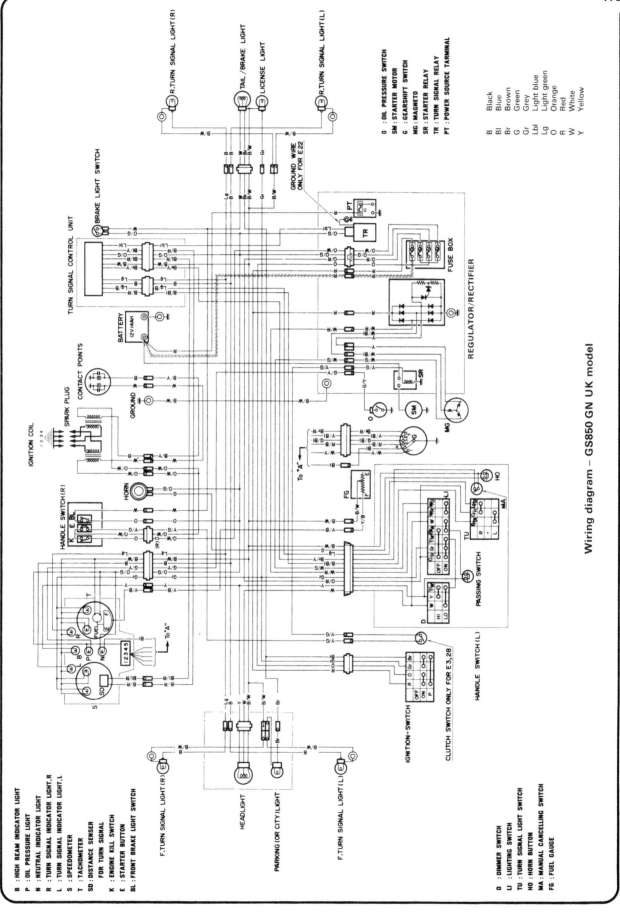

Wiring diagram – GS850 GN UK model

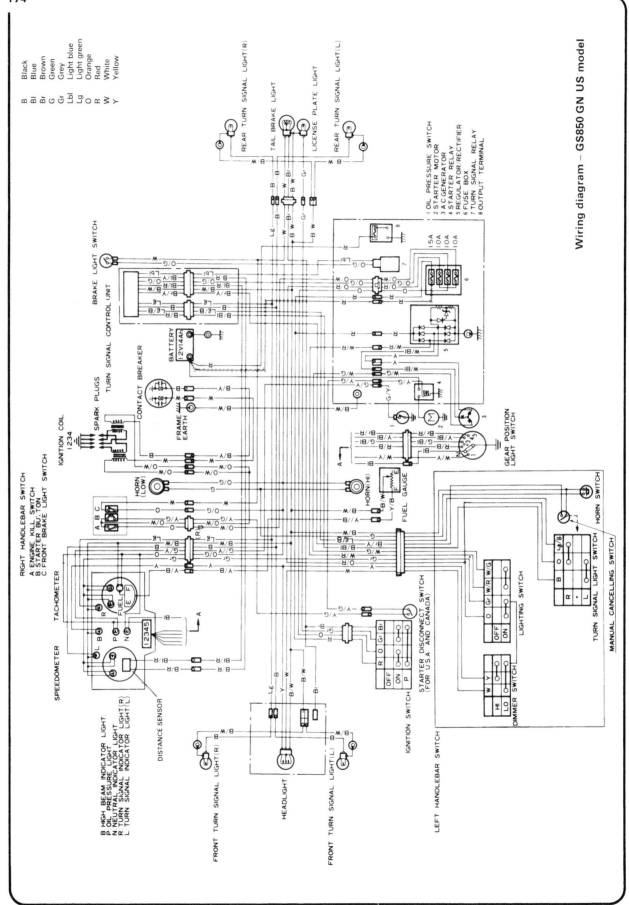

Wiring diagram – GS850 GN US model

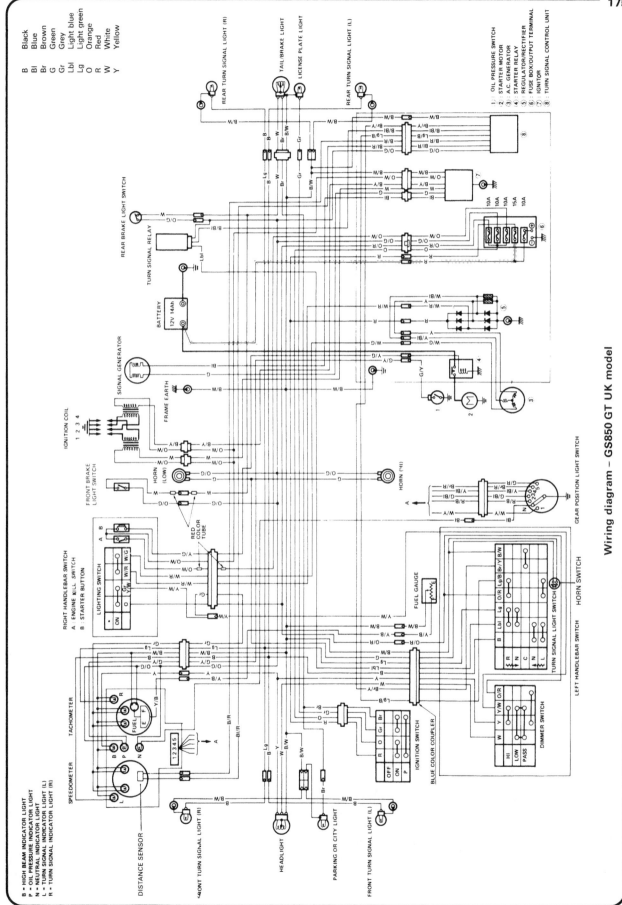

Wiring diagram – GS850 GT UK model

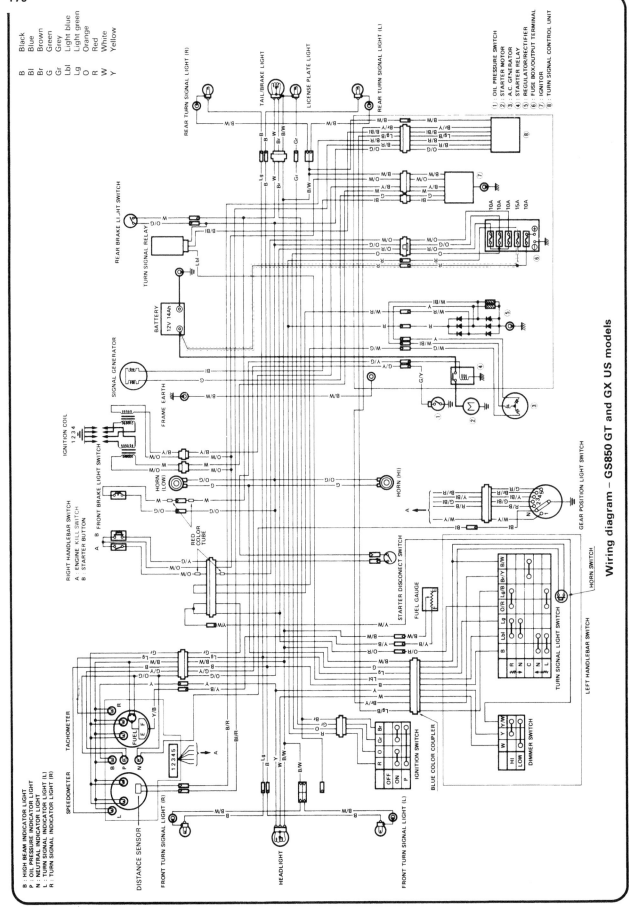

Wiring diagram – GS850 GT and GX US models

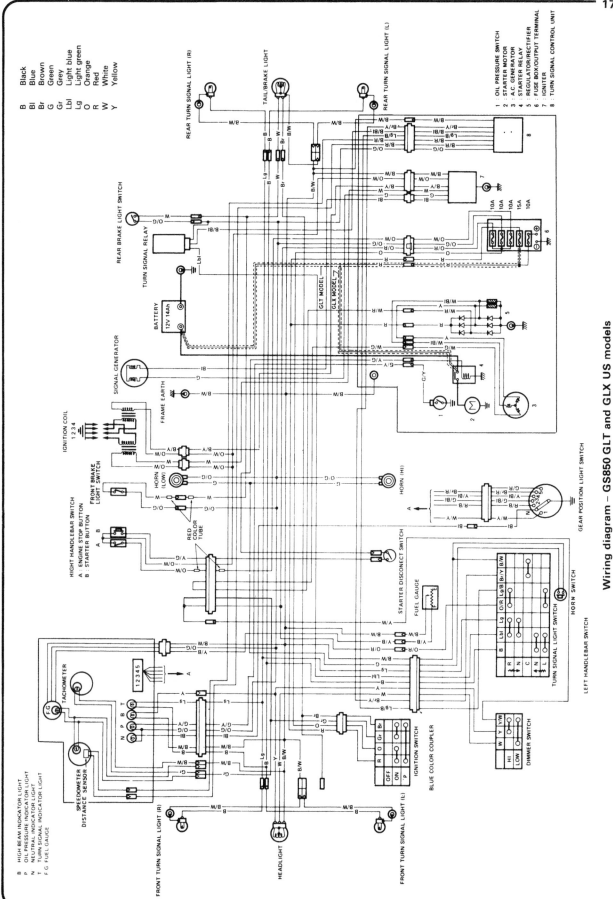

Wiring diagram – GS850 GLT and GLX US models

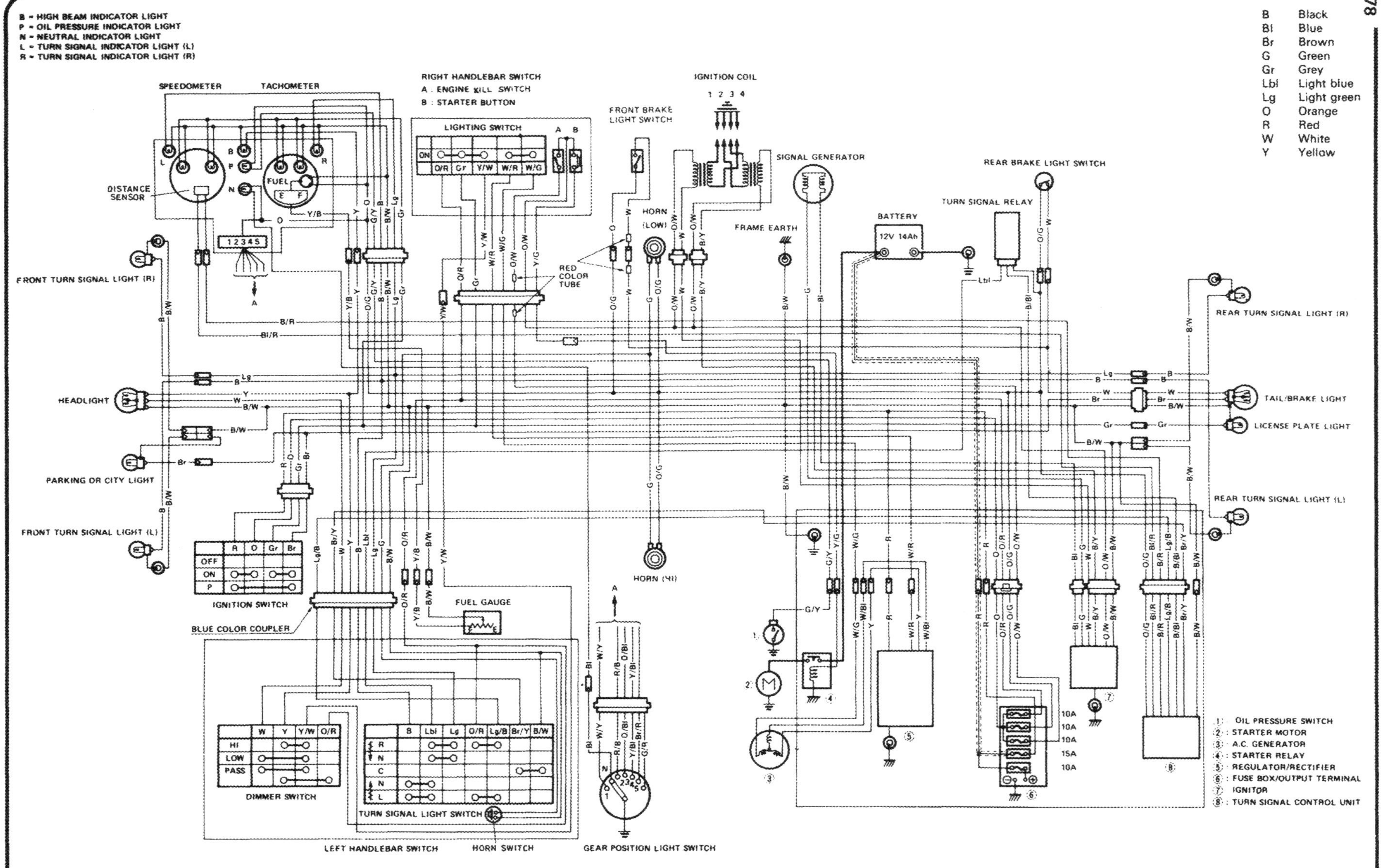

B Black
Bl Blue
Br Brown
G Green
Gr Grey
Lbl Light blue
Lg Light green
O Orange
R Red
W White
Y Yellow

B = HIGH BEAM INDICATOR LIGHT
P = OIL PRESSURE INDICATOR LIGHT
N = NEUTRAL INDICATOR LIGHT
L = TURN SIGNAL INDICATOR LIGHT (L)
R = TURN SIGNAL INDICATOR LIGHT (R)

REAR TURN SIGNAL LIGHT (R)
TAIL/BRAKE LIGHT
LICENSE PLATE LIGHT
REAR TURN SIGNAL LIGHT (L)

1 : OIL PRESSURE SWITCH
2 : STARTER MOTOR
3 : A.C. GENERATOR
4 : STARTER RELAY
5 : REGULATOR/RECTIFIER
6 : FUSE BOX/OUTPUT TERMINAL
7 : IGNITOR
8 : TURN SIGNAL CONTROL UNIT

REAR BRAKE LIGHT SWITCH
TURN SIGNAL RELAY
BATTERY 12V 14Ah
SIGNAL GENERATOR
FRAME EARTH
IGNITION COIL
1 2 3 4

10A 10A 10A 15A 10A

FRONT BRAKE LIGHT SWITCH
HORN (LOW)
HORN (HI)
GEAR POSITION LIGHT SWITCH

RIGHT HANDLEBAR SWITCH
A : ENGINE KILL SWITCH
B : STARTER BUTTON
LIGHTING SWITCH

RED COLOR TUBE

FUEL GAUGE

HORN SWITCH
TURN SIGNAL LIGHT SWITCH
LEFT HANDLEBAR SWITCH

DIMMER SWITCH
HI LOW PASS

IGNITION SWITCH
OFF ON P
BLUE COLOR COUPLER

SPEEDOMETER
TACHOMETER
DISTANCE SENSOR
FRONT TURN SIGNAL LIGHT (R)
HEADLIGHT
PARKING OR CITY LIGHT
FRONT TURN SIGNAL LIGHT (L)

Wiring diagram – GS850 GX UK model

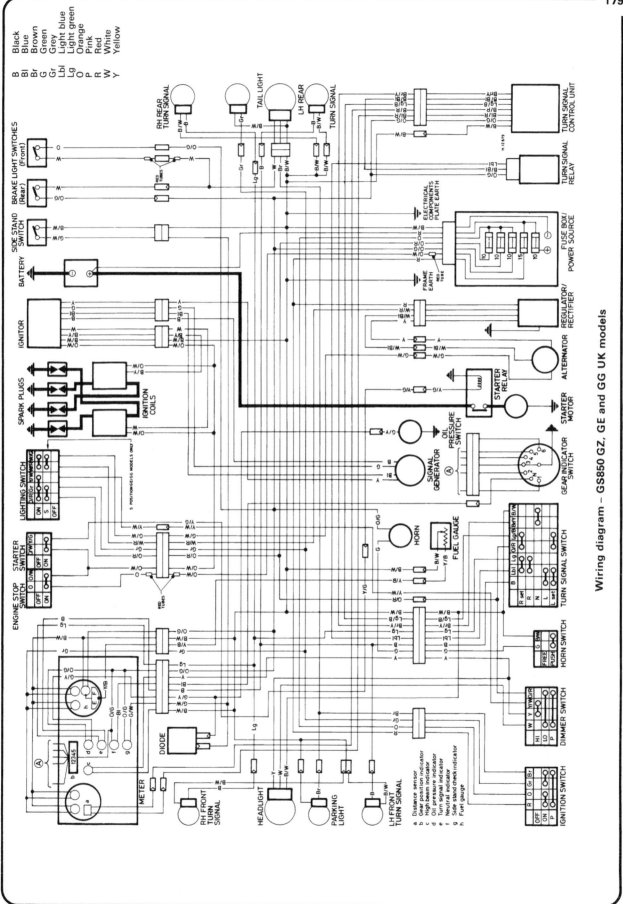

Wiring diagram – GS850 GZ, GE and GG UK models

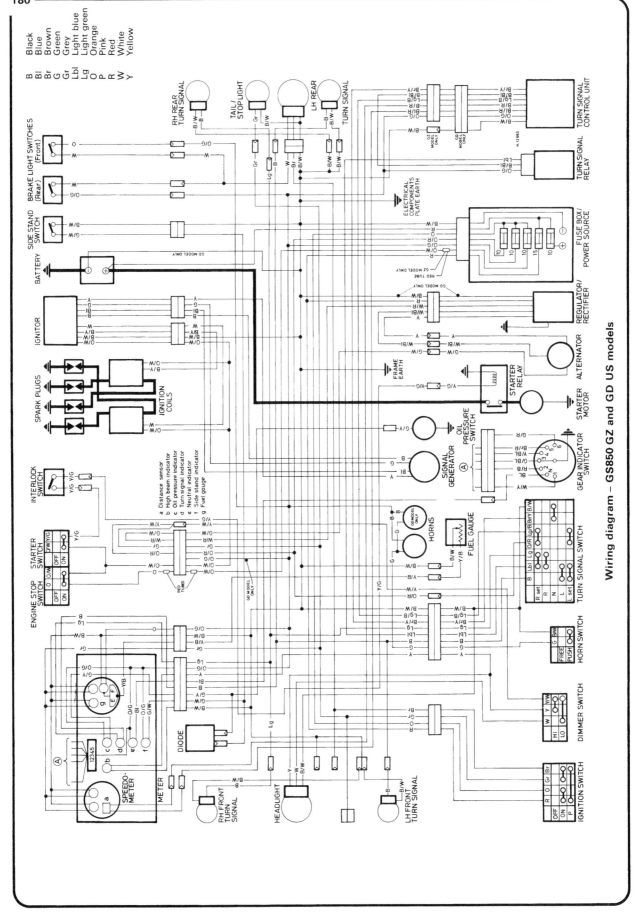

Wiring diagram – GS850 GZ and GD US models

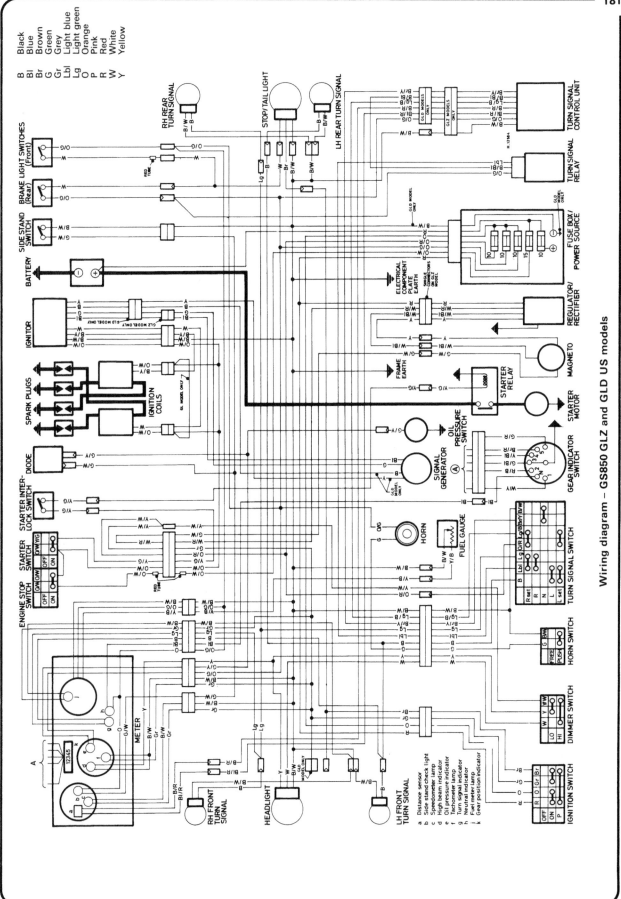

Wiring diagram – GS850 GLZ and GLD US models

Index